CHILD CARE—
WHO CARES?

CHILD CARE WHO CARES?

Foreign and Domestic
Infant and Early Childhood
Development Policies

STUDENT EDITION

EDITED BY PAMELA ROBY

FOREWORD BY SHIRLEY CHISHOLM

BASIC BOOKS, INC., PUBLISHERS
NEW YORK

Copyright © 1973 by Basic Books, Inc.
Introduction to the Student Edition
Copyright © 1975 by Pamela Roby.
Library of Congress Catalog Card Number: 72-89179
International Standard Book Number: 0-465-09526-7
Designed by Vincent Torre
Printed in the United States of America
75 76 77 78 79   10 9 8 7 6 5 4 3 2 1

For My Mother
and Children around the World

Contents

PART III

Cross-National Perspectives on Child Care: A Look Abroad

Foreword

United States Congresswoman

For the past several years Congress has tried to legislate a comprehensive program of child-care services in the United States—a program that would allow parents and community groups to work together to create healthy and full environments for children whose mothers and fathers work or might be able to work were they assured decent child care. As a member of Congress and of the House Education and Labor Committee, I have participated in endless discussion of possible day-care programs and have heard the gamut of clichés that accompany criticism or denial of the need for child care—clichés that were summed up in November 1971 when President Nixon vetoed one legislative plan that could have given the United States a beginning of sound and progressive national policy toward the health of its children and their families.

Child Care—Who Cares?, coming as it does at a time when America still, as Pamela Roby says, has a "nonsystem" of services for young children, will be a valuable reference for those who wish to study child care in depth. The book covers diverse aspects of child-care programing and policy: (1) discussion concerning who would use the service, including infants who, because of mistaken comparisons between institutionalization and child care, have been deprived of child-care programs since the early 1950s; (2) an examination of the private and public services that are currently available in the United States, including a history of past services and perspectives on child care from special interest groups; (3) a discussion of current problems and issues surrounding the provision of more adequate services in the United States; and (4) a thorough analysis of child-care systems abroad, including a description of services in Sweden, Finland, Israel, Hungary, Norway, England, the Soviet Union, and Japan.

Part III, covering current issues, should be of particular interest to members of Congress, directors of child-care centers, and persons involved in making social policy. The chapter on legislation clearly summarizes those problems that have attended many closed-door arguments in Congress; those chapters on community control, governing centers, and cost provide enough data and examples to demonstrate the complexities that have stalled programs and made some of our legislative efforts look like "child's" patchwork. Pamela Roby's summary is excellent reading for anyone who wants an up-to-date review of the child-care crisis in this country by someone whose point of view is humanistic—not mechanistic.

Ms. Roby's message, and one that can be found throughout the book, is that discrimination against women and indifference to the poor are at the center of this country's reluctance to expand its support for child-care programs. Woman's place is in the home—so the attitude goes—and if women *choose* to work, then they'll have to make the necessary arrangements themselves.

As documented in *Child Care—Who Cares?*, this argument is not borne out by the facts. Women don't choose to work. They have to. And "arrangements" don't exist. Three million mothers are rearing their children in fatherless homes. Two out of three (64 percent or 1,920,000) of these mothers are the sole providers for their children. The rest of the women in our thirty-two million strong female work brigade are supporting themselves or together with their husbands are supporting their families. Poor, working poor, lower middle class, middle class—they are all in the same boat. They are, like their husbands, breadwinners. In nearly one-third of our families where both parents work, the husband's income is less than $5,000. As for "arrangements," only 2 percent of our women use group day-care facilities. The rest face a nightmare of hodgepodge arrangements with elderly relatives, a rapid turnover of babysitters, and the bleak custodial parking lots euphemistically called family-care centers.

If you are lucky, a family-care center means that your child will be safe, clean, fed, and lovingly cared for by a gentle soul who likes children. More likely than not, you won't be lucky and the person in charge may be emotionally disturbed, uneducated, alcoholic, so old as to need help herself, or all of the above.

Many countries abroad (See Part IV) have comprehensive day care; by comparison the United States is definitely not a leader in providing quality care. Nor are many of the major political voices at this time focusing on the need to do something for America's children.

Existing programs and many proposed programs emphasize service for the poor—an emphasis that is justified because of "concern" for "limited" funds. Funds in the United States aren't limited—they're just spent in other places. We are the richest nation in the world. We scrimp on programs for people because we choose to spend our money on tanks, guns, missiles, and bombs! We even limit child-care funds for the poor. According to the testimony of Elizabeth Koontz, director of the Women's Bureau:

> The lack of child care services has been the most serious single barrier to the success of the Work Incentive (WIN) program. Care in centers for eligible children is rare and most mothers in the program have been forced to make their own arrangements. These have proved to be haphazard and subject to frequent changes, interruptions, and breakdowns.

There is no question that the solution of the welfare problem and the economic problem in the United States is irrevocably linked with the necessity of providing good and easily accessible day-care services. When we talk about welfare we are talking about AFDC families— mothers and children. Study after study has shown that many welfare mothers want to work, but they are not going to work unless they feel their children are safe and well cared for.

There are so many, many things that can be done in this country and that *ought* to be done. What we need now is the will to carry them out. *Child Care—Who Cares?* is a book that is giving us a much needed and objective push in the right direction.

Introduction to
the Student Edition

PAMELA ROBY

Many professors and students throughout the nation have asked for a paperback edition of *Child Care—Who Cares?* Therefore, I am pleased that Basic Books has decided to publish this student edition which contains the essence of the original volume. It will be especially valuable in courses in the areas of early childhood education, social policy, women's studies, sociology of education, educational administration, social work, and child welfare.

The primary purpose of *Child Care—Who Cares?* is to provide the groundwork for developing child care policies and programs in the United States. The book remains current because little change has occurred with regard to child care policies in the United States over the last several years. The purpose of this new introduction will be to review the recent literature, organizational developments, statistics, and policy and political developments pertaining to child care.

The growth of interest in child care has been reflected in the considerable recent literature and current research on the subject; a bibliography of the literature which has appeared since the first edition of this book went to press is included at the end of this chapter.

The Need for Child Care

The growing interest in child care is the result of an increasing need for it as well as of child care advocates' improved political organization. Despite the declining preschool population, the number of children under age six whose mothers were in the U.S. labor force increased by half a

* I thank Dana Friedman Tracy of the Day Care and Child Development Council of America, Inc. for her assistance in the preparation of this introduction.

million (from 5.6 to 6.1 million) between 1970 and 1974.[1] Currently one-third of all preschool children have mothers in the labor force. Year-round full-time jobs were held by 25 per cent of all mothers with at least one preschooler.[2] Most of the absolute increase from 1970 to 1974 in the number of children with employed mothers occurred in female-headed families and reflected the rising divorce rates in the United States—the highest in history.[3]

Most mothers who worked had to—they were either divorced, widowed, separated, or their husbands were unemployed or earned incomes too low to support a family. The 6 million preschool-aged children of these employed mothers all require child care. In addition, many other pre-schoolers need the group experience which child care provides. In 1972 (the latest available statistics), the United States had a total of 81,286 licensed (or approved) day care centers and family day care homes with a total licensed capacity of slightly over 1 million children—e.g., a capacity which met less than one-sixth of the need for child care.[4]

Organizational and Political Developments

Given the large and growing need for child care, it is fortunate that child care advocates have not only become increasingly well organized and politicized, but have also grown in number. Three national organizations, as well as numerous regional, state, and local groups concerned with child care have been formed since the first edition of this book went to press. They are now working actively on behalf of children. Three new national organizations are listed at the conclusion of this introduction. These are in addition to the older associations listed in the Appendix.

Many individual child care centers have also joined together to form local or regional associations to coordinate fund raising, purchasing, train-ing, and other activities. The Wichita Child Day Care Association (WCDCA) which grew from three centers in 1972 to 91 in 1975 is one example. It "offers its participating members technical assistance on administrative problems and proposals as well as nutritional, nursing, social work, speech therapy, and parental supportive services; inservice training for staff; a centralized library of publications and film strips about child care; and a legislative clearing house." In addition, the WCDCA staff aggressively seeks out and compiles a wide range of information concerning funding for its participating programs.[5]

At least as important, persons concerned with child care have learned much about the legislative process and have become more politically sophisticated. The child care forces, for example, sent 250,000 letters to the U.S. Congress and the Department of Health, Education, and Wel-fare commenting on the child-care regulations proposed by the Department in 1973 which would have prevented many children from receiving day care. As a consequence, Congress placed a moratorium on the proposed HEW regulations and developed Title XX of the Social Security Act discussed below. The passage of Title XX in December 1974 showed that people concerned with children were willing to work legislatively for child care.

The people who were involved in the Title XX struggle learned much about the political process and helped others become more politically aware. They learned the necessity of developing legislative proposals around which all child care advocates could unite and of involving as many segments of the child care community as possible in this process. They also learned the necessity of political education. Recently, for example, The National Association for the Education of Young Children and other organizations concerned with the well-being of young children have developed legislative task forces and have made legislative resolutions at their annual conventions.

Additionally, over the last four years "child care" has come to mean much more than institutional day care. Today it refers to family child care; part-day or "drop-by care;" preventative health care and medical care for children; parent cooperatives; associations of family day care and child care centers; and nutritional, counseling, funding, teacher training, parent training and other support services. This broadened definition of child care has been an important source of political unity.

Policy Developments

Two important federal acts have been passed and one major bill has been introduced concerning child care over the last several years. The acts—Title XX of the Social Security Act and the Tax Reduction Act, and the Child and Family Services Bill are described below. In addition, numerous states have enacted child care legislation of their own. For information concerning the latter developments, contact the Education Commission of the States listed in Appendix C.

TITLE XX OF THE SOCIAL SECURITY ACT

Title XX of the Social Services Amendments of 1974 (Public Law 93-647) took effect October 1, 1975. Title XX provides up to $2.5 billion to be allocated to the states according to population for, among other social services, "child care services, protective services for children, services for children in foster care, services related to the management and maintenance of the home, and appropriate combinations of services designed to meet the special needs of children." [Sec. 2002(a) (1)]. Under the law, the Federal government provides 75 percent of the funding for the services and the states provide 25 percent in matching funds. The services provided must be related to at least one of the following five goals: (1) economic self-support; (2) self-sufficiency; (3) preventing or remedying neglect; abuse, or exploitation of children, or preserving, rehabilitating, or reuniting families; (4) preventing or reducing inappropriate institutional care; or (5) securing referral or admission to institutional care when other forms of care are not appropriate, or providing services to individuals in institutions (Sec. 2001). Under Title XX, each state has much discretion in defining services, determining eligibility, and deciding which agencies to fund. The Governor of each state must specify the state agency to administer Title XX. This agency has complete administrative authority with the exception that it must follow the regulations of the law.

Among the regulations of Title XX are eligibility requirements. States may include in a Title XX social-services program any person who is a member of a family with a monthly gross income which does not exceed 115 percent of the median income for a family of four in that state, adjusted for family size. In addition, in-home day care must meet standards established by the state. Day care outside the home must meet the Federal Interagency Day Care Requirements of 1968 (FIDCR), with the following exceptions: (1) the FIDC requirements for educational services are optional, and (2) the Secretary of the Department of Health, Education, and Welfare must write staffing standards for the care of children under age 3 (there are none under the FIDCR). Under the FIDC Requirements, day care programs must have an advisory committee that includes parents.

Under Title XX, states must publish a comprehensive annual services plan and make it "generally available to the public at least 90 days before the beginning of a services program year." At least 45 days after the proposed plan is published and prior to the start of the program year, the state must publish the plan in final form, with an explanation of any changes and the reasons therefore. Any amendments to the final plan must be published in proposed form, with a 30-day public comment period (Sec. 2004).

A benefit of the new law is that it provides for the first time funding for child care for the working poor without the necessity of their being labeled a "former or potential welfare recipient." For anyone with a gross monthly income above 80 percent of the state median family income or 100 percent of the national median family income, whichever is lower, the law requires states to charge a fee, reasonably related to income [Sec. 2002(a) (6) (B)]. The Secretary of HEW may, but need not, prohibit fees for welfare recipients and others below the 80 percent level [Sec. 2002(a) (5)]. Other benefits of the new law include considerable expansion of the list of services which states may provide under the federal-state 75-25 percent reimbursable match; the possibility and encouragement of reasonable integration of services; funding for the training and retraining of day care staff, family day care providers, and others associated with child care services; and funding for local planning and evaluation services for child care and other programs.[6]

THE TAX REDUCTION ACT

In 1975 President Ford signed the Tax Reduction Act which allows families with incomes up to $35,000 to make a tax deduction of $4,800 a year for child care. Families with incomes from $35,000 to $44,600 may deduct decreasing amounts for child care on a sliding scale basis.[7]

THE CHILD AND FAMILY SERVICES BILL

Representatives Brademas, Heckler, Bell, and Mink introduced the Child and Family Services Act of 1975 (H.R. 2966) on February 6, 1975. They were supported by eighty-six cosponsors. The next day, Senators Walter Mondale and Jacob Javits, along with 24 cosponsors, introduced the Senate version of the Bill (Sec. 626).

The purpose of the Child and Family Services Bill (CFS) is to help families better meet their need for quality, family-oriented, preschool programs for young children whose mothers are working, or who because of inadequate resources are denied adequate health care, nutrition, or educational opportunities. Toward this end, the bill authorizes the expenditure of $1.85 billion for training, planning, technical assistance, and program operation over a three-year period. The bill, if enacted, would also provide financial assistance for the construction or acquisition of nonprofit child care facilities if they are necessary for the establishment of adequate services. Financial assistance would also be provided for the training of individuals employed or preparing for employment in child care work including volunteers, professional, and nonprofessional personnel.

The bill also provides for an Office of Child and Family Services which would be maintained within the Department of Health, Education, and Welfare, would have a Director appointed by the President, and would assume the responsibilities of the present Office of Child Development as well as administer the CFS Act. A federal Child and Family Services Coordinating Council would be established consisting of the directors of the Office of Child and Family Services and other federal social services agencies. Its purpose would be to maximize available resources and prevent duplication of services. If the bill is enacted, the Secretary of the Department of Health, Education, and Welfare would be charged with making an annual evaluation of federal involvement in CFS services including an analysis of expenditures; description of available programs; and an analysis of the effectiveness of programs, and the extent to which preschool, minority, and economically disadvantaged children and their parents have participated in the programs. The Office of CFS would be responsible for regular and periodic monitoring of programs and for providing a trained staff to do this.

Prime sponsors, under the bill, could not be profit-making agencies but could be a state, locality, or combination of localities. Prime sponsors would be responsible for coordinating the delivery system; establishing and maintaining a Child and Family Services Council; approving the CFS plan, policies, procedures, and budget; and overseeing an annual and ongoing evaluation of the services under its jurisdiction. At least 50 percent of the members would be parents of children in the program and one-third economically disadvantaged. The CFS Council would consist of ten or more members and would be responsible for approving the CFS plan, policies, procedures, budget, and evaluation.

The major issues which have arisen concerning the Bill are the role of public schools and profit-making child care providers in the delivery system, state and local rather than federal control of funds, the funding level, and the amount of parent participation.[8] Although the CFS Bill has much support, these issues will have to be resolved for it to have a chance of passage.

The Future

Hopefully the Child and Family Services Bill will be enacted for it would bring this nation one step closer to quality child care for our children. Whether or not the bill passes, however, we will have a considerable way to go before we can rest knowing that we have quality child care. Our efforts must continue until we have child care policies which include not only fine child care centers for all children who can benefit from them, but also good preventative and remedial health care, nutrition, education, recreation, and family-support services so that all children may be assured of a healthy emotional environment, good parent-child, sibling-and-familial interaction, adequate housing, and a decent, secure family income.[9]

NOTES

1. Between March 1970 and March 1974, the population of children under six years declined in the United States from 19.6 to 18.5 million. Elizabeth Waldman, "Special Labor Force Report: Children of Working Mothers, March 1974," *Monthly Labor Review*, Vol. 98, No. 1, January 1975, p. 64; cf., Elizabeth Waldman and Robert Whitmore, "Children of Working Mothers, March 1973," *Monthly Labor Review*, Vol. 97, No. 5, May 1974.

2. *Ibid.*

3. *Ibid.*, p. 66, the rate was 4.7 (per 1,000 of the population) in 1974, as compared with the previous all-time high of 4.3 in the immediate post-World-War-II period and rates of 2.1 to 2.6 from 1950 to 1967, and 2.9 to 4.0 from 1968 to 1972.

4. As of 1972 there were 20,319 licensed day-care centers and 60,967 licensed family day care homes. The day care centers had an approved capacity of 805,361 children and the family day care homes a capacity of 215,841. "New National Child Care Statistics," *Voice for Children*, Vol. 7, No. 11, December 1974, p. 3.

5. "Wichita: The How (and Why) of Becoming a Child Care System," *Voice for Children*, Vol. 8, No. 3, March 1975, pp. 1–2.

6. Cf., "Title XX: Making It Work For Child Care," *Voice for Children*, Vol. 8, No. 3, March 1975, p. 6; Judy Assmus, "Title XX: A Summary of The New Social Services Law," *Voice for Children*, Vol. 8, No. 2, February 1975, pp. 2–3. For more information concerning Title XX and child care, contact the Day Care and Child Development Council of America, Inc., 10012 14th Street, NW, Washington, D.C., 20005.

7. In March 1975, Senator John V. Tunney introduced legislation to allow working mothers to deduct child care costs as business deductions on their income-tax returns. Representatives James Corman and Bella Abzug introduced a similar bill in the House. The Tax Reduction Act described above permits working mothers to deduct child care costs only as a personal expense. According to Tunney, that classification rules out the tax benefit for the 68 percent of the nation's families with yearly earnings of less than $10,000 who use the standard tax form which doesn't permit itemized deductions. Tunney's amendment would cover widowers and divorced men with children as well as working mothers. *Women Today*, Vol. V, No. 7, March 31, 1975, p. 37, and interview with Robert McNair, Administrative Assistant to Senator Tunney.

8. Cf., Toye L. Lewis, "The Pros and Cons of the Child and Family Services Act of 1975," *National Council for Black Child Development Newsletter*, Vol. I, No. 3, April 1975, pp. 4, 6; Dana Friedman, "The Issues and The Testimony," *Voice for Children*, Vol. 8, No. 4, April 1975, pp. 7–9; "Federal Day Care Program Urged to Fill Growing Need," *AFL-CIO News*, Vol. XX, No. 19, May 10, 1975, p. 2; Testimonies to the Joint Hearing of the House Select Subcommittee on Education, the Senate Subcommittee on Children and Youth, and the Senate Subcommittee on Employment, Poverty, and Migratory Labor, Washington, D.C., February, March, June 1975.

9. Elsewhere, I have discussed the need for social policies which support "shared parenting" or fathers playing a larger parenting role. Cf., Pamela A. Roby, "Shared Parenting: Perspectives From Other Nations," *The School Review*, Vol. 83, No. 3, May 1975, pp. 415–431; Henry Biller, *Father, Child, and Sex Role: Paternal Determinants of Personality Development*, Lexington, Massachusetts: Heath Lexington Books, 1971; Biller, *Paternal Deprivation*, Lexington, Massachusetts: Heath Lexington Books, 1974; Biller, *Father Power*, New York: David McKay Company, 1974; David B. Lynn, *The Father: His Role in Child Development*, Monterey, California: Brooks/Cole Publishing Company, 1974.

NEW NATIONAL ORGANIZATIONS CONCERNED WITH CHILD CARE

National Association for Child Development and Education, 500 12th Street S.W., Suite 810, Washington, D.C. 20024. Represents all of the licensed proprietary day-care center operators in the United States.

National Council for Black Child Development (NCBCD), 490 L'Enfant Plaza East, S.W., Suite 3204, Washington, D.C. 20024. Broadly based national umbrella organization of interested professional caucuses, community organizations, day care and child development agencies, interested individuals and groups with advocates for black children and youth, their families and the communities of which they are a part; and is supportive of all positive efforts directed toward the improvement of the welfare of black children and youth.

National Council of Organizations for Children and Youth (NCOCY), 1910 K Street, N.W., Washington, D.C. 20006. Umbrella organization of 200 national, state and other organizations concerned with day care, foster care, and the treatment of juvenile delinquents.

RECENT LITERATURE ON CHILD CARE

Asmussen, Patricia D. *Simplified Recipes for Day Care Centers,* Boston: Cahners Books, 1974.

Auerbach, Stevanne. *Child Care: A Comprehensive Guide,* Vol. I, New York: Behavioral Publications, Inc., 1975.

Baker, Rebecca, et al. *A Lap to Sit on and Much More: Helps for Day-Care Workers,* Washington, D.C.: Association of Childhood Education International, 1971.

Bane, Mary Jo. "Who Cares About Child Care?" *Working Papers For a New Society,* Vol. 2, No. 1, Spring, 1974.

Blackstone, Tessa. *Education and Day Care For Young Children in Need: The American Experience,* New York: International Publications Service, 1974.

Boocock, Sarane. "A Cross-Cultural Analysis of the Child Care System," in Lillian Katz (ed.), *Current Topics in Early Childhood,* New York: Academic Press, forthcoming.

———. "Children and Society," in Arlene Skolnick (ed.), *Children in Society,* Boston: Little, Brown and Company, forthcoming.

Bourne, Patricia G. "The Three Faces of Day Care," in Louise K. Howe (ed.), *The Future of the Family,* New York: Simon and Schuster, 1972.

———. *The Unconglomerated Agglomerate: Child Care and the Public Sector,* Berkeley: Department of City and Regional Planning, University of California, Berkeley, Ph.D. dissertation, 1974.

Breibart, Vicki. *The Day Care Book: How, What and Why of Rational Day Care,* New York: Alfred Knopf, 1972.

Bremner, Robert H. (ed.). *Children and Youth in America: A Documentary History,* Vol. III, 1933–1970, Cambridge: Harvard University Press, 1974.

Callahan, Sidney. *The Working Mother,* New York: Macmillan Company, 1971.

Costin, Lela B. *Child Welfare: Policies and Practice,* New York: McGraw-Hill, 1972.

Day Care and Child Development Council of America, Inc., *Resources for Child Care: Complete Catalog of Publications for 1975,* Washington, D.C.: Day Care and Child Development Council of America, Inc., 1975.

Denzin, Norman K. *Children and Their Caretakers,* New Brunswick, New Jersey: Transaction, 1973.

Drach, Howard. "The Politics of Child Care in the 1940's," *Science and Society,* Vol. 38, No. 2, Summer, 1974.

Ehrensaft, Diane. *Socialization of Sex Roles in a Day Care Center,* Ann Arbor: Department of Psychology, University of Michigan, Ph.D. dissertation, 1974.

Ellis, Katherine and Rosalind Petchesky. "Children of the Corporate Dream: An Analysis of Day Care as a Political Issue Under Capitalism," *Socialist Revolution,* No. 12, Nov.-Dec., 1972.

Evans, E. Belle and George E. Saia. *Day Care for Infants: The Case for Infant Daycare,* Boston: Beacon, 1972.

Fein, Greta and Alison Clarke-Stewart. *Day Care in Context,* New York: Wiley, 1973.

Fink, Stevanne Auerbach. *Parents and Child Care,* San Francisco: Far West Laboratory for Educational Research and Development, 1973.

Ford, Sally. *Guidelines for Day Care Service,* New York: Child Welfare League of America, 1972.

Goldsmith, Cornelia. *Better Day Care For the Young Child: The Story of Day Care in New York City,* Washington, D.C.: National Association for the Education of Young Children, 1972.

Gordon, Ira J. *Early Childhood Education: The Seventy-First Yearbook of the National Society for the Study of Education,* Part II, Chicago: University of Chicago Press, 1972.

Griffin, Al. *How to Start and Operate a Day Care Home,* Chicago: Henry Regnery Company, 1974.

Harvard Education Review. "Special Issues on Children's Rights," Vol. 44, January 1974 to February 1974.

Hoover, Mary B. *Home Study Course in Day Care Services,* New York: Child Welfare League of America, 1973.

Hunt, Joseph W. and Eleanor D. Craig. "Should We Provide More Government Funding for Day Care?" *Public Policy,* Vol. XX, No. 4, Fall, 1972.

Joffe, Carole. "Child Care: Destroying the Family or Strengthening It?" in Louise K. Howe (ed.), *The Future of the Family,* New York: Simon and Schuster, 1972.

————. *Marginal Professionals and Their Clients: The Case of Child Care,* Berkeley: Department of Sociology, University of California, Berkeley, Ph.D. dissertation, 1974.

Lord, Catherine and Renee Watkins. *Storefront Day Care Centers: The Radical Berlin Experiment,* 1973.

Marcus, Joseph (ed.). *Growing up in Groups: The Russian Day Care Center and the Israeli Kibbutz,* New York: Gordon and Breach, 1973.

Prescott, Elizabeth, Cynthia Milich, and Elizabeth Jones. *The Politics of Day Care,* Vol. I, Washington, D.C.: National Association for the Education of Young Children, 1972.

Prescott, Elizabeth Jones, and Sybil Kritchevsky. *Day Care as a Child-Rearing Environment,* Vol. II, Washington, D.C.: National Association for the Education of Young Children, 1972.

Robinson, Halbert B., et al. *Early Child Care in the United States of America,* New York: Gordon and Breach, 1973.

Robinson, Halbert B. and Nancy M. Robinson (eds.). "International Monograph Series on Early Child Care," monographs on child care in Cuba, France, Great Britain, Hungary, India, Israel, Poland, Sweden, Switzerland, the Soviet Union, and Yugoslavia. New York: Gordon and Breach, 1973.

Rothman, Sheila. "Other People's Children: The Day Care Experience in America," *The Public Interest,* No. 30, Winter, 1973.

Shane, Harold G. and Robert H. Anderson (eds.). *As The Twig is Bent: Readings in Early Childhood Education,* Boston: Houghton Mifflin, 1971.

Steiner, Gilbert. "Day Care Centers: Hype or Hope?" in Helen Z. Lopata (ed.), *Marriages and Families,* New York: D. Van Nostrand, 1972.

————. "Child Care: New Social Policy Battleground," *The Brookings Bulletin,* Vol. 9, No. 1, Winter, 1972.

Steinfels, Margaret O'Brien. *Who's Minding the Children? The History and Politics of Day Care in America,* New York: Simon and Schuster, 1973.

United States Office of Child Development, Department of Health, Education, and Welfare, *Infant Care,* New York: Arco Publishing Company, 1975.

————. *Research, Demonstration and Evaluation Studies,* Fiscal Year, 1973, 1974, Washington, D.C.: U.S. Office of Child Development, 1973, 1974.

United States Women's Bureau, Department of Labor, *Day Care Facts,* Washington, D.C.: U.S. Women's Bureau, Department of Labor, 1973.

Weaver, Kitty. *Lenin's Grandchildren,* New York: Simon and Schuster, 1971.

Young, Dennis R. and Richard R. Nelson. *Public Policy for Day Care of Young Children: Organization, Finance and Planning,* Lexington, Massachusetts: Heath Lexington Books, 1973.

Preface to the First Edition

Several converging stimuli prompted this book. First was my own concern about how many of my friends and I could combine the rearing of happy, healthy children with our career commitments. This personal concern was joined by my continuing interest in the education and development of young children, an interest that has been with me since long before my course work and student teaching in early childhood education. The women's movement then reinforced my interest in child-rearing and child development. It not only stressed parents' need for good group child-care programs but also began child-care programs voluntarily staffed by men as well as women, and childless persons as well as parents. These and other newly founded centers for young children clearly showed that good group experiences contributed much to young children's happiness and development.

Prompted by this ever increasing interest in infant and early childhood development programs, I began to ask friends and acquaintances, especially those from other nations and those involved in American preschool programs, what they thought and knew about child care. What were preschool programs and policies like in Germany? in Sweden? in Japan? What was happening in America's own varied preschool centers? What should be happening? Probably more than anything else, the responses and interest of these friends led to this book. I visited their own or their friends' nurseries, kindergartens, family day-care centers, and child-development programs throughout many parts of the United States, Finland, Denmark, Norway, Sweden, and Israel.

Child Care—Who Cares?, written following these visits, examines and evaluates diverse infant and early childhood development policies in the United States, Finland, Sweden, Norway, Hungary, England, the Soviet Union, Israel, and Japan. The book's policy perspective grew out of my own background in social policy analysis and the feeling on the part of myself and others that amid much discussion of what could or should happen within a preschool room, crucial social policy questions were being neglected. Which children are to have the benefit of attending child-development programs? How are the programs to be financed?

Who is to staff them? How are they to be governed? What role are parents to have in child-care programs? Are child-care resources to be distributed to insure greater or lesser inequality? What would be the optimum conditions for involvement of minorities and women in child-development programs?

Another defining feature of the book is that it is about *preschool* child-development programs. This feature too reflects my own background and training. I hope someone else will write the sorely needed volume on after-school programs.

Finally, because child-development considerations cannot be confined within the bounds of child-care programs, the contributors to this volume examine not only child-care centers but also the distribution of national resources, prenatal and postnatal health and nutrition programs for mothers and children, maternity leave policies, birth allowances, children's or family allowances and other income maintenance policies, and parent education programs. Children are seen as whole beings whose development is vitally linked to their families' and communities' conditions of living. Neither the establishment of good child-development centers nor the construction of any other single social program are regarded as panaceas for the problems of children. *The full potential for children of good child-development programs such as those recommended in this book will be realized only when racism, sexism, and class inequalities no longer shape the lives of citizens and limit their options.* It is hoped that *Child Care—Who Cares?* will prove useful to local, state, and federal policy-makers, advocates for children, and early childhood educators and program directors by raising issues for discussion, research, planning, and action pertaining to child-development programs, as well as to professors in schools of education and social work who are teaching courses on early childhood.

The work for this book was greatly facilitated by a Ford Foundation Research and Travel Grant, and by the financial support of the Center of Manpower Policy Studies, George Washington University, and the Social Policy Study Program, Brandeis University, which is funded by the U.S. Children's Bureau. I am grateful to Sar A. Levitan and David G. Gil, directors of these programs, for granting me the opportunity to concentrate on editing this book.

I have benefited from the advice and criticism of several colleagues, especially S. M. Miller, Jessie Bernard, Maria Carvainis, Wilma Scott Heide, Kay Cassell, and Martin Rein. Virginia Kerr, Susan Stein, Mary Potter Rowe, Evelyn Moore, and Adele Simmons read large portions of the volume and made substantial suggestions. Astrid Brekken helped tremendously by translating and editing the chapter on Norwegian child care. Among the many persons who furnished special calculations and other data for the book were Charles Gershenson, Joan Hutchinson, Suzanne Woolsey, Preston Lee, Gordon Hurd and Mike Marquardt, all of the U.S. Department of Health, Education and Welfare; Richard

Emery of the U.S. Office of Management and Budget; Larry Feldman and Rita Alfer, formerly of the Day Care and Child Development Council of America; Gwenn Morgan of the Massachusetts Office of Planning and Evaluation; Mats Hellström, Member of Parliament and Chairperson of the Royal Commission on Child Care Facilities, Sweden; Ingrid Arvidsson, Cultural Attaché of the Swedish Embassy, Washington, D.C.; Elina Haavio-Mannila of the University of Helsinki; and Lise Heber Østlyngen of the University of Oslo.

I also wish to thank the staff of the many children's centers I visited; Nava and Shlomo Shmneli with whom I stayed on Kibbutz Tzora in Israel; my parents, Clark D. and Marianne G. Roby, for a constant stream of clippings and articles on child-care policies; and Iris Diamond, Jackie Potter, and Virginia Normann who did both research for and typing of many chapters.

This book is dedicated to children around the world and to my mother, Marianna Gilman Roby, who has devoted much of her life to her own children and to those in her kindergartens.

PAMELA ROBY

Cambridge, Mass.
1973

The Authors

RIVKA BAR-YOSEF-WEISS, former kibbutz member, is presently Chairperson of the Department of Sociology at the Hebrew University in Jerusalem. At the Hebrew University she also studied psychology and education. She is engaged in research focused on the sociology of the family, especially in modern societies, and on social welfare planning. Dr. Bar-Yosef-Weiss is also Co-director of the Israeli Center for Policy Studies and Consultant to the Bureau of Planning and Research of the National Insurance Institute of the Israeli Ministry of Labor.

THE BLACK CHILD DEVELOPMENT INSTITUTE, understanding that the future of the Black race depends upon the strength of its children, has as one of its primary objectives development of a technical assistance model that will enable Black community organizations to establish comprehensive child-development programs that will serve as a catalyst for the advancement of the total community. Concurrent with this effort is an attempt to devise administrative, curriculum, staff, and economic development models that will meet the needs of Black children, their families, and their communities. It is through the creation of these models that the Institute predicates its commitment to the restoration of this vital institution in our community. BCDI's Technical Assistance Project was designed to offer technical assistance to a small number of child-development programs in grass-roots Black communities across the country. In addition BCDI hopes to provide a strong Black presence in the child-development field. Through its newsletter, bulletins, and position papers it tries to keep communities abreast of developments in the field and provides them with information on how to best run quality child-development centers.

BETTYE M. CALDWELL is Professor of Elementary Education at the University of Arkansas and Director of the Center for Early Development and Education in Little Rock, Arkansas. The Center is a comprehensive educational child-care facility for infants, preschoolers, and elementary school children. Dr. Caldwell is also editor of Volume 3 of the *Review of Child Development Research* and was until recently the editor of *Child Development*. She serves on many national committees concerned with developing or evaluating programs for children, including the recent White House Conference on Children and Youth. Her own needs for day care are over, as her thirteen-year-old twins now not only take care of themselves but have become a valuable neighborhood resource for day care!

ANTONIA HANDLER CHAYES, former Dean of Jackson College, Tufts University, is presently Associate Professor of Political Science at Tufts and a lecturer at Boston University's Law School. She was formerly Director of Urban Development for Action for Boston Community Development (ABCD, Boston's antipoverty agency), Coordinator for Boston's Model Cities Program, and Social Science Advisor on the staff of the National Institute of Mental Health's Office of Planning.

IRJA ESKOLA is a member of Helsinki's Commission on Children's Day Care and Research Director of the recent Survey of Child-Care Needs in Helsinki. She has taught in the Department of Social Policy of Helsinki University and has authored numerous articles on children's day care, the status of women, and social policy in Finland.

SUSAN FERGE is a Research Sociologist at the Sociological Research Institute of the Hungarian Academy of Sciences and a Reader in Sociology at the Central University of Budapest. She is now conducting research on educational social structures and problems of social policy. She has authored three books and numerous articles in these and other areas.

ELIZABETH HAGEN is a member of the Board of Directors of the Princeton University/National Organization for Women Day Nursery, which she describes in her chapter. She is also doing research on sex stereotypes in children's school textbooks and working actively with the women's liberation movement in the Princeton area.

JAMES L. HYMES, JR., has been Professor of Early Childhood at the University of Maryland, at George Peabody College for Teachers, and at the State University of New York. Dr. Hymes is past President of the National Association for the Education of Young Children and past Vice-President of the Association for Childhood Education International. He was a member of the National Planning Committee for Project Head Start and during World War II was Director of the Kaiser Child Service Department in Portland, Oregon—the world's largest day-care center for young children of working mothers. He has authored numerous books for teachers, parents, and children, including *Understanding Your Child* (1952), *Before the Child Reads* (1958), *The Child under Six* (1963), *Oodles of Noodles* (1964), and *Teaching the Child under Six* (1968).

MARIANNE KARRE is a Swedish journalist who has specialized in preschool education and the facilities offered for this by the community. She is now serving as a consultant to the Royal Commission for Child Supervision.

VIRGINIA KERR is a coordinator, writer, and researcher in the national office of the National Women's Political Caucus in Washington, D.C. She has worked as a Field Representative of the Day Care and Child Development Council of America; a Program Development Officer at Teachers College, Columbia University; and has taught in Glasgow, Scotland, and New York City.

PAMELA ROBY is Associate Professor in the Departments of Sociology and Community Studies at the University of California, Santa Cruz. She is trained in early childhood education as well as sociology and taught in the Jefferson County, Colorado, public schools. Dr. Roby has also taught at Brandeis, George Washington and New York universities and has worked as a Research Sociologist at the Center for Manpower Policy Studies, George Washington University; the Center for Human Relations, New York University; the Russell Sage Foundation; and the Youth Development Center, Syracuse University. Co-author with S. M. Miller of *The Future of Inequality* (Basic Books, 1970), she is an associate on the National Manpower Policy Task Force, a consultant to the U.S. Department of Labor on Welfare Reform Research and Child Care, an advisor for the Cambridge Institute Child Care Resource Center, a Consultant to the White House Conference on Children's Forum on Day Care, and active in Sociologists for Women in Society. She has written extensively on issues in education, the status of women, social welfare, and social policy; and during 1970 and 1971 met with child-development specialists and visited child-development centers in Finland, Sweden, Norway, Denmark, and Israel.

ADELE SIMMONS, formerly Dean of Jackson College and an Assistant Professor of History at Tufts University, is now Dean of Students, Princeton University. Dean Simmons was instrumental in the establishment of the Tufts day-care center. She is presently preparing a Twentieth Century Fund report on problems of working women.

PART I

Who Needs Child Care?

1

CHILD CARE—WHAT AND WHY?

PAMELA ROBY

After years of neglect the United States is now turning its attention to infant and early childhood development programs. Child care is attracting increasing attention because of the American economy's growing dependence upon working mothers, the rise in single-parent families, and parents' increasing recognition that their young children benefit from good group experiences and early education.

In March 1970 there were 21,332,000 children under six years old in the United States.[1] Six *million* of these children had working mothers, an increase of two *million* children from 1960. The percentage of mothers with children under six who were in the labor force increased from 19 to 30 percent during the 1960s; the proportion of Black and single mothers who were employed was much higher.[2] Two and a half million, or over 12 percent of all U.S. children under six, lived in families headed by a female. Three and a half million, or over 17 percent of all U.S. children under six, lived in families with incomes falling under the U.S. Office of Economic Opportunity's poverty line of $3,885 for a family of four;[3] 45 percent of these low-income children lived in female-headed families. An additional 2.2 million children under age six lived in families with incomes under $6,960, the "low family budget" of the Bureau of Labor Statistics.[4]

Most mothers work, the U.S. Department of Labor has shown, because of financial need. The husbands of 43 percent of the mothers with children under six who worked had incomes under $7,000; these mothers obviously worked out of financial necessity. The husbands of another 34 percent of working mothers with preschool children had incomes between $7,000 and $10,000; these mothers worked to ease their families' fairly tight financial situation.[5]

Another smaller but rapidly growing group of women view the opportunity for mothers of young children to work or study—an opportunity that child-care programs provide—as a basic human right. Young mothers today are, after all, better educated than ever before, and they want to use their education. Twenty-three percent of mothers with preschool children who worked had husbands who earned over $10,000.[6] These women probably worked for a combination of reasons: they liked their jobs and wanted to continue to develop their skills, and both their families' financial situation and their own status within their families were improved by the money they earned. One mother recently reported to *Time* magazine, "When I put my check on the dining-room table, I get respect; I do not get that ironing shirts." [7] Mothers of many more children need to work for financial reasons but are unable to do so because of the lack of good child-development facilities.

[In addition to the rapidly growing demand for child-care facilities for children of mothers who work or study, the nation is beginning to realize that preschool education and development programs are needed for the physical, emotional, and intellectual development of all its youngsters. Educators and psychologists are finding that children's development during these early years significantly affects their later ability to learn and grow.] Although child-care programs are most often discussed in connection with low-income children, national statistics tell us that children of high-income or highly educated parents are more likely to be enrolled in preschool programs than children of low-income or less educated parents. While Head Start programs are hailed as giving low-income children a fair start in life, much larger numbers of already advantaged children attend public preschool programs provided in suburban areas or private programs paid for by their parents. In 1970, 38 percent, or over four million, of all U.S. three, four, and five year olds were enrolled in preschool programs.[8] The percentage of children enrolled increased by age: 13 percent of all three years olds, 28 percent of all four year olds, and 69 percent of all five year olds were enrolled in a preschool program with a developmental component.[9] Of these children 31 percent of the three year olds, 23 percent of the four year olds, and 12 percent of the five year olds were enrolled in a full-day program.[10]

The percentage of children attending preschool programs was directly correlated with their parents' income and education. Forty-eight percent of the three to five year olds with family incomes of $10,000 and over, as compared with only 24 percent of those with family incomes under $3,000, attended a preschool program; 53 percent of those in families headed by a college graduate, as compared with only 24 percent of those in families headed by a person with less than eight years of schooling, attended preschool programs.[11] In addition to telling us that we are severely remiss in providing preschool programs for poor children, these data suggest that as parental incomes and educations continue to rise, the demand for child-development centers will grow even greater.

How do we now care for children of working mothers? Most children of working mothers are cared for in their own homes. According to the Westinghouse-Westat study supported by the Office of Economic Opportunity, 55 percent of all children in out-of-home, full-day child care are cared for in family day-care homes, that is, by an adult who cares for a group of children in her or his own home on a regular basis; the other 45 percent are cared for in child-care centers. More than one-fifth of the children in family day-care homes are under age two.

Although a smaller percentage of children of working mothers are cared for in child-care centers than in family day-care homes or their own homes, the study shows that more of the mothers whose children are in centers are well satisfied with their child-care arrangements. The least satisfactory arrangements, according to the working mothers interviewed, are those involving a sibling or nonrelative caring for the child in the home or in an unlicensed family day-care home.[12] Less than 2 percent of the estimated 450,000 family day-care homes are licensed as compared with almost 90 percent of the child-care centers, even though government funds pay the fees for many of the children cared for in the homes.[13] About 575,000 children, of whom 24,000 are estimated to be under two years old, received full-day care in an estimated 17,500 child-care centers. Sixty percent of these centers are proprietary, and proprietary centers care for about half the children enrolled in centers. People working in day-care centers are, on the whole, very poorly paid. In 1970 teachers received a median ten-month salary of $3,580 and teachers' aides received $3,000. Most worked at salary levels under the national minimum wage and many were below the poverty line.[14]

The National Council of Jewish Women took a closer look at the child-care arrangements and "nonarrangements" in which the children of working mothers spend so much of their time. Of the 431 child-care centers that Council members visited across the nation, only about a quarter offered developmental care including educational, nutritional, and health services. Hundreds of children in these centers seemed to be getting an exciting, happy start in life. All too many other young children—thousands—were found to be grossly neglected, living ten hours a day on their own, staying day after day at their mothers' places of work because no other arrangements could be made for them, or somehow surviving in child-care centers or family day-care homes of such poor quality that they may suffer lasting damage. The following were among the Council members' reports:

Peter, age three, gets his own lunch every day. He has to. No one else is home. . . . He eats what he can reach and what his still uncoordinated hands can concoct if he can get the refrigerator or cabinet doors open. . . . Peter is anything but alone in his plight. The (Chicago) City Welfare Department estimates that 700 children less than six years of age are left alone each day without any formal supervision when their mothers leave for work.

This is an abominable center. It was very crowded. In charge were several untrained high school girls. No adults present. No decent toys. Rat holes clearly visible. To keep discipline, the children were not allowed to talk.

This center . . . was absolutely filthy . . . broken equipment . . . broken windows . . . 2 children, aged ten and twelve, in charge. The kitchen was very dirty.[15]

These children's mothers did not hate them. Rather, forced to work with adequate child care unavailable, the mothers had no choice but to leave their children in undesirable situations. The child-care shortage increased between 1965 and 1971. Over that time, while the number of working mothers with children under six increased from 3.7 million to 4.6 million, the number of full-time, year-round household workers decreased by 200,000.[16] In the early 1970s, federal funds for day care provided services for fewer than 5 percent of the children of low-income families most in need, and all licensed nonprofit and proprietary centers in the United States had an enrollment capacity in 1970 of about 625,000 children including those of school age. Many of these centers provided only part-day care; most of the "full-day" centers were not open long enough to meet the needs of women who had to get to work very early or who worked late or on weekends.[17]

At long last, local, state, and federal governments, churches, women's liberation groups, universities, businesses, hospitals, and other private organizations are beginning to grapple with America's great need for good group child care. Hundreds of state and federal legislators are introducing child-care bills. Architects are drawing up a variety of nursery school plans.

Before nationwide child-care policies become institutionalized, we would do well to consider seriously the alternatives that are available to this country. We are at the moment of decision—the pressing need for child care does not allow us to pause—but as we formulate policies that will affect millions of children, it is our responsibility to survey the knowledge that already exists concerning group child care so that we do not make mistakes that others have already learned to avoid. In the area of child welfare, as in that of old age, the United States lags far behind many developed and developing nations. *While it is the misfortune of our children today that the nation has a severe shortage of good child-development centers, it is our obligation to draw upon the experience of other nations as well as our own past experience to make excellent child-development programs the fortune of all our children tomorrow.* This volume is devoted to this end.

The chapters that follow were written specifically for this book by persons with a wide variety of backgrounds and experiences with child care. Their writings examine child care from the perspective of social policy and programing as contrasted with the pedagogical perspective.[18] "Child care," the focus of all the chapters, refers to all forms of group care for healthy infants and preschool children who live with their parents for part of each day; it includes all forms of child-center care (creches, infant day-care centers, nurseries, kindergartens, day-care centers, and child-development programs) as well as family day care and care of children in their own homes when it is an aspect of social policy, for example, when in-home care is partially or fully financed or supervised by a governmental or voluntary organization's program.[19]

Child Care—Who Cares? is composed of four parts, each of which examines a different aspect of child-care policy and programing. Part I explores the need for child care in highly industrialized nations such as the United States. Proposals for the creation and the expansion of comprehensive child-development programs are examined and suggested in Part II. Part III critically reviews U.S. experiences with child care including: the historical development of nurseries, kindergartens, and day-care centers; recent federal, state, university, company, and women's liberation child-care center programs and policies; and a federally sponsored, municipally administered family day-care program. In Part IV, early childhood experts of Sweden, Finland, Hungary, Norway, England and Wales, Japan, and Israel describe and analyze their nations' child-development policies including prenatal and postnatal health care, children's allowances, maternity leaves, nutrition programs, and child care.

The contributors to *Child Care—Who Cares?* view questions of child development as part of larger issues concerning the treatment of women, the poor, minorities, and children. These questions involve problems of priorities, values, and choice in the area of social policy.

The welfare of millions of young children is a great challenge and responsibility. It is hoped that this book will help its readers and those with whom they work construct comprehensive child-development programs that will meet the needs of all children.

NOTES

1. The number of children born annually has declined slightly over recent years. In March 1970, 3,704,000 children were under one year old; 3,566,000 were one year old; 3,482,000 were two years old; 3,481,000 were three years old; 3,576,000 were four years old; and 3,723,000 were five years old. U.S. Office of Economic Opportunity, *Current Population Survey March 1970 Supplement* (Washington D.C., 1971), p. 49. The 1970 U.S. census also taken in March showed 21,017,690 children under age six. U.S. *Summary of General Population Characteristics, Advanced Reports* (Washington D.C.: U.S. Government Printing Office, 1971).

2. Bureau of Labor Statistics, U.S. Department of Labor, *Survey Shows Substantial Rise in Number of Young Children with Working Mothers* (Washington D.C.: Office of Information, U.S. Department of Labor, May 26, 1971), p. 1. Cf. Elizabeth Waldman and Anne M. Young, "Marital and Family Characteristics of Workers, March 1970," *Monthly Labor Review* 94 (March 1971): 46; Elizabeth Waldman and Kathryn R. Gover, "Children of Women in the Labor Force," *Monthly Labor Review* 94 (July 1971), p. 19.

3. U.S. Office of Economic Opportunity, *op. cit.*, table 5, p. 49. Cf. Robert L. Stein, "The Economic Status of Families Headed by Women," *Monthly Labor Review* 93 (December 1970), p. 3.

4. Michael Marquardt, Office of Statistics, Office of Child Development, U.S. Department of Health, Education, and Welfare, statistical worksheet; Bureau of Labor Statistics, *op. cit.*, Table 1, p. 2.

5. Waldman and Young, *op. cit.*, Table K, p. A–18. A 1965 Department of Labor survey of six million working mothers found that 87 percent said they worked for financial reasons. Seth Low and Pearl G. Spindler, *Child Care Arrangements of Working Mothers in the United States*, Children's Bureau, no. 461 (Washington D.C.: U.S. Government Printing Office, 1968), p. 9.

6. Waldman and Young, *op. cit.*, Table K, p. A–18.

7. Editors, "A Woman's Place Is on the Job," *Time* 98, no. 4:56.

8. Elementary and Secondary Surveys Branch, National Center for Educational Statistics, U.S. Department of Health, Education, and Welfare, *Pre-primary Enroll-ment October 1970*, by Gordon E. Hurd (Washington D.C.: U.S. Government Printing Office, 1971). The data are based upon information gathered by the Bureau of the Census in its October 1970 Current Population Survey. The series of preprimary enrollment surveys were begun in 1964. "Preschool" as defined in the survey ques-tionnaire was to include only child-care programs with an instructional component; children in solely custodial care were not counted as being enrolled in preschool. The data refer only to three to five year olds enrolled in preschool programs at the time of the survey; additional three to five year olds were enrolled in programs dur-ing other months of the year.

9. *Ibid.*

10. *Ibid.*

11. *Ibid.*

12. Westinghouse Learning Corporation and Westat Research, Inc., *Day Care Survey 1970*, report to the Office of Economic Opportunity pursuant to contract OEO BOO–5160, April 1971, pp. vii. The Westat survey does not tell us whether mothers were more satisfied with *licensed* family day care than with informal, *unli-censed* family day-care arrangements. Since 98 percent of the family day-care homes as compared to only 10 percent of the child-care centers were unlicensed, we need to know the degree to which mothers are satisfied with licensed family day-care homes as compared to licensed child-care centers to validly base policy on mothers' preferences. It should also be noted that the child-care preferences mothers have re-ported to surveyors have often been conditioned by what the mothers know they can realistically afford to pay for child care.

13. *Ibid.*, p. vii.

14. *Ibid.*, pp. vii, ix, 64.

15. Mary Dublin Keyserling, *Windows on Day Care: A Report on the Findings of Members of the National Council of Jewish Women* (New York: National Council of Jewish Women, 1972), pp. 13, 14. In 1966 a Women's Bureau-Children's Bureau study reported that 18,000 children under the age of six were latchkey kids on their own. Not only has this number probably grown with the increased day-care short-age, but it was likely to have been an underestimate originally since many mothers probably will not tell a census taker that they are unable to make any arrangements for their children. Low and Spindler, *op. cit.*

16. *Ibid.*, p. 12.

17. *Ibid.*, pp. 2–4.

18. Detailed discussions concerning Montessori, Piaget, Froebel and specific find-ings relating to cognitive development in young children and what should be in-cluded in preschool curricula are contained in other volumes and therefore do not need to be examined here. See, for example, Laura L. Dittmann, ed., *Early Child Care: The New Perspectives* (New York: Atherton Press, 1968); Philip Lichtenberg and Dolores G. Norton, *Cognitive and Mental Development in the First Five Years* (Chevey Chase, Md.: National Institute of Mental Health, 1970); Joint Commission on the Mental Health of Children, *Crisis in Child Mental Health: Challenge for the 1970's* (New York: Harper & Row, 1969), pp. 137–164, 313–329; Annie L. Butler, *Current Research in Early Childhood Education* (Washington D.C.: American Asso-ciation of Elementary-Kindergarten-Nursery Educators, 1970); Sandra Byford Wake, Dorothy O'Connell, and Charlene Brash, *Research Relating to Children*, ERIC Clearinghouse on Early Childhood Education, Bulletin 27 (Washington D.C.: U.S. Government Printing Office, 1971); Robert D. Hess and Roberta Meyer Bear, eds.,

Early Education: A Comprehensive Evaluation of Current Theory, Research and Practice (Chicago: Aldine Publishing Co., 1968); Joe L. Frost, ed., *Early Childhood Education Rediscovered* (New York: Holt, Rinehart and Winston, 1968); Martin L. Hoffman and Lois Wladis Hoffman, *Review of Child Development Research*, 2 vols. (New York: Russell Sage Foundation, 1964); Bettye Caldwell, *Educational Child Care for Infants and Young Children* (New York: Holt, Rinehart and Winston, forthcoming); Edith H. Grotberg, ed., *Day Care: Resources for Decisions* (Washington D.C.: U.S. Government Printing Office, 1971); R. K. Parker, ed., *Conceptualizations of Pre-school Curricula* (Boston: Allyn and Bacon, 1971); J. McVicker Hunt, *Intelligence and Experience* (New York: Ronald Press, 1961); Judith Chapman, Joyce Lazar, and Edith Grotberg, *A Review of the Present Status and Future Needs in Day Care Research* (Washington D.C.: U.S. Office of Child Development, 1971). For a guide to how to plan, develop, and operate a day-care center, see E. Belle Evans, Beth Shub, and Marlene Weinstein, *Day Care* (Boston: Beacon Press, 1971); for a guide to choosing nursery schools, see Jean Curtis, *A Parents' Guide to Nursery Schools* (New York: Random House, 1971).

19. In addition to these terms, it should be noted that in the very confusing lingo of preschool policy, "twenty-four hour child care" refers to child-care programs in which care is *available* around the clock rather than to centers, as some have mistakenly thought, that care for an individual child around the clock ("twenty-four hour care" is most important, as will be discussed later, to mothers who must work night shifts, who pursue evening studies, or who are confronted with an emergency late at night; in addition, the 1970 Westat survey (p. 109 shows that nearly 60 percent of all working or student mothers who use day-care centers leave their homes before 8 A.M. and/or return after 5 P.M.). "Universal child care" refers to programs that are available to all children within the "universe"—state or nation—described; and "comprehensive child care" to programs that not only watch over and educate the children involved but also provide health care and other social services as they are needed. In addition to all the types of child-care centers described above, "child care" will also be used in this volume to refer to "family day care", that is, part- or full-day care of infants and young children provided in their own home by an individual usually not related to them. In this book "child care" will not be used to refer to parents' care of their own children or to care provided in orphanages or other "total institutions."

2

CHILDREN'S NEED FOR

CHILD CARE

JAMES L. HYMES, JR.

Specialists in early childhood development regard a good group experience as very desirable for the young child. They regard three, four, and five year olds in particular as very ready to go to school for part of the day. They feel quite sure that children benefit from the experience.

Some who are not professionally involved with the early childhood ages regard group living for young children primarily as a possibility for certain selected "special" children. They might favor, for example, a group for a three, four, or five year old who is an only child. Or for some isolated youngster who has no companions his own age with whom to play. They often support early group experiences for children with some physical handicap (the blind, the hard-of-hearing, the crippled, the cerebral palsied, youngsters with speech handicaps). They might support the idea of groups for young mentally retarded children and for youngsters with some emotional disturbance. And, of course, there has been wide support for early schooling for those special youngsters disadvantaged by the impact of poverty. Early childhood professionals agree that nursery schools, kindergartens, children's centers—whatever name you want to give to groups for young children—can be useful indeed for such special youngsters and—they add—especially useful for all young children.

Almost everyone agrees, of course, that when a mother is working or ill, or for any other reason not available to her three, four, or five year old, a good children's group may well be the answer. Early childhood specialists point out that the child gets into the group because of a special family situation, but once he is in it, a good group can serve the child's needs well.

The experience of those in early childhood development is that the "extra-special" quality—only child, handicapped child, poor child, working mother—is not crucial. The fact that the child is three, four, or five is special enough. His developmental age means he is ready for an early childhood group.

Many people are offended and worried by this whole notion. Their backgrounds lead them to think that school begins at age six. When they hear of groups for younger children they assume that their mothers must be trying to get rid of the youngsters.

Beyond any doubt mothers with children in early childhood groups have a wide variety of motives. Some do reject their youngsters. Some send them unthinkingly simply because today early schooling seems the "in" thing to do. But through the years the best early childhood groups have been sustained by conscientious and committed parents, not by the uncaring or "bandwagon" ones. By sending their child to early school these parents have not felt they were shirking their responsibilities, but rather exercising them. They have carefully sought out a good group, not so they would have less to do at home but so the group would support and supplement what they do for their child at home.

Critics worry about the future of the family and fear that early childhood groups will weaken home life. The conscientious parents would reply that having their child in a group has helped to make them wiser, more skilled, and better parents, not worse. Critics fear that sending children to school so early means that the state is taking over the child's life. This anxiety is almost incomprehensible to parents whose children are in a good group. In their eyes the early childhood group is an enrichment of the child's life, an added advantage, like having access to a community museum, a topflight library, or a wonderful hospital or a park. It doesn't take the place of the home, but extends what even the best home is able to do for its children.

For many—both those troubled by and pleased with this view—groups for children under age six sound like a recent innovation, even a slightly daring step, one still in the experimental stage. Some dates may help to put this in perspective. The first private kindergarten in the United States was established in 1856. In 1873 the first kindergartens were opened as a part of a public-school system in St. Louis, Missouri. The first private nursery schools in the United States were established in 1921 and 1922. In 1925 the first nursery school was opened in a public school, the Franklin School of Chicago. The oldest professional organization concerned with early childhood, today called the American Association of Elementary-Kindergarten-Nursery Educators, was founded in 1884. The largest early childhood professional organization in the country today was founded in 1892, then known as the International Kindergarten Union and today as the Association for Childhood Education International. The second largest, and the one specializing most in the ages under six, is the National Association for the Education of Young Children, begun in 1925. By 1926 there were eight major research centers in this country emphasizing the study of young children: the Child Welfare Research Station of the University of Iowa (1917); Yale University's Psycho-Clinic (1919); the Merrill-Palmer Institute in Detroit (1922); Harvard University's Psycho-Educational Clinic (1922); the Child Development Institute of Columbia University (1924); the Child Development and Family Life Department of Cornell University (1925); the Institute of Child Welfare of the University of Minnesota (1925); the Institute of Child Welfare of the University of California at Berkeley (1926). So much research about young children had been undertaken, so much experimental and demonstration work done, that in 1929 the National Society for the Study of Education devoted its twenty-eighth yearbook, an 875-page volume, to *Preschool and Parental Education.*

America has had a history of groups for young children sponsored by churches, by labor unions, by industries. We have had private profit-making groups and private nonprofit groups. There have been groups of young children in high schools and on liberal arts college campuses to contribute to preparental education. There have been groups of children on other university campuses to contribute to teacher education, medical education, nursing education, religious education.

More will constantly be found out about young children. More will be found out about how best to meet their needs in a group. *But* America has been at it a long time. We know ever so much right now.

The central question, of course, is the age of the children. Many people have only one image in their minds of three and four year olds in particular but often also of five year olds: They really are "just babies." . . . They are so young. . . . They need their mothers. Such people find it hard to fit this image into any picture of school as they know it.

Out of the long history of research, study, experimentation and demonstration comes a different image of the child. Although, three, four, and five year olds are young and do need their parents, they need more than their parents. They need other children and a good group experience.

Early childhood school may seem like a cold, harsh term, especially to those who have few fond memories of their own school days. Friedrich Froebel (1782–1852), regarded as the father of the kindergarten, coined that wonderfully expressive term: *children's garden.* This term is usually reserved for five-year-old groups (although we do have a few free public four-year-old kindergartens too). But "kindergarten" in its literal sense would make a very apt term for all groups for children under six: nursery school, day nursery, child-care center, day-care center. A children's garden. A special place. A spot of earth planned, equipped, sized, and staffed expressly for young children, with only their particular needs and peculiar requirements in mind.

We talk today of planned communities for our aging and for the retired. We talk of planned communities in our suburbs and of planned communities for recreation in our mountains and beside our lakes. Call it kindergarten, call it school, call it child care, or any other name. A good group for young children is a planned community that focuses specifically on what a young child is like and on what will help him function at his very best.

For one thing young children are very specifically body conscious. Such a short time ago in their brief living, they were immobile. Then in a rush of development they crawled; they pulled themselves up; they stood alone; they took their first steps. This was a very heady experience, to come so suddenly into such a wealth of new powers.

Parents at home with their children say their youngsters wear them out; the children are constantly on the go, bent on using and perfecting their newly found bodily skills. In one research study a trained athlete tried to duplicate the movements of a two year old, which had been captured on film. The athlete became exhausted as parents do.

The young child needs some spot in his life where his absorption in locomotion, in climbing, in balancing, in swinging, in speeding can be appreciated and encouraged, not merely tolerated, squelched, or prohibited. A good group for young children has ample outdoor space, and the space is richly equipped with wagons and tricycles, with ropes and ladders and challenging climbing apparatus with boards and boxes and barrels. The outdoor time is not a skimpy recess, a grudging break from the real job of school, but a central facet of the program. This safe, supervised, but generous time for free-and-easy movement with daring adventuring is one of the reasons why a young child in a group can feel so completely that *this is living*.

Young children are also exceedingly social. This quality is one of the least appreciated aspects of early development. Young children are egocentric to be sure; they are very dependent on mothering adults; they are tentative and shy in their first branchings out—these are all accurate descriptions. But in addition quite early in life—before age two—children grow into a very selective awareness of other young children. At the beginning they like to watch them, to be near them. Then quickly by late age two, by age three and four, young children have bona fide friends and want to do things with their friends.

Almost everyone knows how much friends mean to preadolescents and to adolescents. Young children are every bit as social. Their life has a pleasing lilt and a zestful lift when they can be with their friends. One specialist in the field, asked to identify the greatest hazard facing young children today in our city centers and in our suburbs, replied "loneliness." We know how dreary the lonely life is for the aged. It is as dreary for the very young.

The good group for young children brings friends into their lives. The terms "early schooling," "early childhood education," or "nursery school" may suggest a quiet, shushed place for sitting children. Much of later schooling is like that. But good early child care is a place where the noise of children is a healthy sound, not an irritation. Youngsters talk and giggle and laugh and whisper and sometimes even shout. A good group has equipment carefully chosen to bring children together: blocks, a workbench, a sandbox, and a dress-up and make-believe corner. A good group is never one group; it instantaneously subdivides into "buzz sessions" and stays that way, so children are face-to-face with their friends.

Young children have still another major characteristic served well by a good group: they are deeply curious, driven by a hunger to know more about their world. Parents at home very early become aware of this burning curiosity. The child's questions—What? Where? Why? How? Who?—are both pleasing and wearing. The child's urge to see, to touch, to taste, to try out leads parent after parent to say proudly, but with fatigue: "He's into everything." A good group tries to provoke this sparkling sense of wonder and to nourish the drive to discover. At its best this is done in a very informal way, so informal that the child thinks of what is going on merely as play and many a casual visitor assumes that all the children are doing is playing. The children do not sit at desks in rows. There is no blackboard, no textbooks. There are obviously no lectures. The whole group is brought together as seldom as possible, but there is a great deal of face-to-face, person-to-person talking between the teacher and one or two or three children.

There will be provocative, informative pictures on the walls. There will be storytimes galore, using excellent books, the best fiction and nonfiction, the best poetry and prose for the age. There will be display tables where one may find almost anything under the sun that could lead a child to wonder and to ask. There will be animals in the room and in the outdoor area. And almost always very good music facilities: an excellent phonograph, a piano perhaps, the teacher's own instrument, a guitar, or maybe an autoharp. Magnets, magnifying glasses, microscopes, scales, rulers, abacuses are all standard equipment.

Usually some central activity is planned that will involve many in the group: cooking a stew perhaps or baking cookies. Frequently a visitor has been invited to come to the group—the firemen and their truck have almost become standard. And quite often the group goes out from their home base to see some of the nearby world with their own two eyes. Small wonder the young child comes home from a good group each day feeling a little more skilled, a little wiser, better informed, and more understanding. When it has been done well, the child is never sated but always eager for tomorrow when he can find out still more.

Perhaps the most distinctive characteristic of the early years is the child's highly unusual imagination. As children move on into the school years and into preadolescence, they become more reality-oriented; imaginations wane. But for a stretch of time—especially when they are three, four, and five—make-believe reaches a peak. This playful, fantasy quality permeates almost every minute of the child's day.

Adults are so far removed from this period developmentally that it is difficult for some to be sympathetic with young children's play. But this deep immersion in pretending and make-believe is no waste. It serves very basic and significant developmental purposes at this time in life. Imagination is the young child's first thinking. He comes up with an idea: "This must be our space ship. . . . You must be the passenger." At this stage the idea does not have to be a useful one—that is a later development. Now what counts is that the child is active, taking the initiative, thinking, planning.

Imagination is also the young child's way of sorting out all the inputs he has been receiving; it is a process of clarifying ideas and impressions. At older ages we mull over in our minds what has happened to us. Young children play out what they have experienced.

Imagination plays a vital emotional role too. Youngsters use their play to make themselves feel very important—"I'm the boss. Show me your ticket"—and to make themselves feel small and cared for—"I must be sick and you must be my mommy and you must feed me." They use their play like a very handy, do-it-yourself emotional spigot, one they can turn on to start a flow of the feelings they most need at the moment.

This highly important imagination flourishes under the conditions a group makes possible. The pretending goes best when there are friends to share it with. It feeds on real-life experiences: the trips, activities, visitors, pictures, and stories, the reality the child has known, become the base for his making-believe.

Most important the group provides the unstructured, free-to-use-as-you-will materials that encourage dramatic play: the paint, clay, sand, blocks, dolls, dress-up clothes, play stove, tables, baby carriage, dishes, telephone, cash register, toy cars and boats. The list of equipment specifically designed to foster this healthy child-imagination is almost endless. And many other pieces of equipment, ostensibly designed for other purposes, are used to this end. It is the rare tricycle, for example, that has not been a horse, motorcycle, car, ambulance, truck, fire engine, and spaceship.

Homes can, of course, also provide the outdoor space and equipment, the companionship, the intellectual challenge, the materials, and the encouragement for imagination. Many parents, knowing the deep value of these to their young children, work very hard to bring these about, frequently building cooperative arrangements with other parents. It does take work, it does take planning, it does take time. A good group is by no means the only way these basic needs of the young child can be met. But it is the simplest way. By pooling its resources, a community or a group of parents can usually provide more for all of its children than any one family alone can provide for its own child.

And there is one last characteristic of great importance to the young child that the home alone does have trouble taking into account: the growing independence of the child. For all of their smallness, youngness, tenderness, softness, dependence, and their fundamental need for loving parents, young children have a strong urge, a steadily expanding one, to feel big and to be separate, distinct people. Parents early become aware of this drive: when their youngster insists on feeding and dressing himself, when he seemingly says "No" to every request, when he won't come when he is called. The young child must feel safe, but simultaneously he must feel independent. This basic drive makes the very act of going to school pleasing to the child. Three, four, and five year olds are ready for some separation from their parents. They thrive on having a world of their own, knowing all the while the safe feeling that home and family are not too far away in distance or in time.

The transition from safety to independence must be made sensitively. A young child may well have hesitant moments when he first begins school; the wise group doesn't rush matters and isn't impatient with the child. But when a youngster feels safe in his new setting and realizes that he hasn't given up the safety of his old setting, life becomes even more pleasing.

It is important to recognize also that the relatively brief separation is pleasing and beneficial for the mother too. Like her child a mother may also have some initial moments of anguish at the time of parting. But mothers almost always report in the end how helpful it is to them to have some free time (and peace of mind, knowing their child is safe). The time for uninterrupted housework, for friends, for a job, or just a breather lets mothers come back to their children refreshed, more patient, better mothers for the brief separation.

A good group for children makes mothers (and sometimes fathers) better parents in still another way. Inevitably the group becomes a parent education center. There are group meetings about children and individual parent-teacher conferences. Parents daily see their child in comparison with others his age. This usually is a source of reassurance because conscientious parents are at times hard on their children, expecting too much. But sometimes, if the child actually has some physical, emotional, or mental difficulty, parents seeing their child with others his age may be spurred to get skilled help in solving whatever is the trouble.

Some groups—the cooperative nursery school in particular—require parents to take turns working in the group as assistants to the teacher. This firsthand experience, along with all the reading, meetings, conferences that accompany it, is tremendously beneficial for parents in building better understanding of their children and better skills in working with them.

The benefits to the young child (and to his parents) of a good group can be very great indeed. The group facilitates a youngster's living his days with great satisfaction and fulfillment, satisfying the child's basic needs for healthy body use, for social challenge, for the input of impressions and information, for the expression of his quite special imagination, and for bigness and independence.

In the course of this the child obviously learns countless lessons that help to make him a more civilized member of society: the innumerable aspects of give-and-take between people; the innumerable aspects of a child's relationship to things; the innumerable aspects of health and safety. (One could never detail all a teacher teaches, all a child learns from his own experience, all his age mates teach him.) And the child, without his knowing it, has stuffed a cultural knapsack that he carries with him on into his tomorrows: of words learned, stories heard, people met, songs sung, sights seen, colors noted, comparisons made, discoveries made. Groups for young children use a different terminology than do higher levels of education, but the young child in his group encounters almost every subject matter, from A to Z, arithmetic to zoology. And in a painless but personal and very meaningful way.

Three, four, and five year olds are very ready to go to school for part of the day, and they benefit by the experience. There is, however, one proviso: the children are ready *if* the school is ready for them. There is no magic in "just any old group." Three, four, and five year olds have a particular claim for high quality because this is their first schooling, and because these are especially vulnerable, impressionable years in their development. The magic only happens when a group is right for the child.

The group must be the right size for the age. Young children can be hurt by crowds. Right class size is no more than about twelve three year olds, about sixteen four year olds, about twenty five year olds. The teacher must be professionally trained, a specialist in teaching young children, and paid a professional's wage. Many assume that "anyone can teach little children." In fact this is the hardest teaching, calling for great knowledge and for the very greatest sensitivity. And no matter how few children, there must always be more than one adult. This need not necessarily be expensive—the aide can be a volunteer or a parent. But for the safety of the children, the richness of the program, and individualization, there must be an aide.

Groups for young children need space, outdoors as well as indoors. And they need a wealth of equipment. Some equipment to be sure can be homemade, but the highly trained teacher teaches through the equipment and through what the equipment leads children into doing. There cannot be great economy here. Without a wealth of materials, supplies, and equipment, much of the other expenditures will be wasted.

Unfortunately our society has not always been willing to pay what a good group costs. Good groups for young children cost a great deal of money. We are accustomed to hearing the statement about college-level education: "Tuition pays only about one-third the cost." We make the false assumption that little children mean little costs. Overly large class size, untrained teachers, limited space, meager equipment have kept many children now in groups from the gains that group living could have brought them.

There is also widespread misunderstanding among parents, untrained teachers, and parts of the general public about the purpose of groups for young children. The image of regular school—desks, chairs, quiet, workbooks, primers, a teacher up front talking—is very clear in people's minds. Carrying this image down into the younger years, however, can be harmful. Too many groups for young children are schools where quiet children sit and learn passivity. The one clear goal of these groups is to "get the child ready" for whatever lies ahead: in nursery school to get him ready for kindergarten; in kindergarten to get him ready for first grade.

The getting-them-ready groups bring young children very little of what the children want: the body use, the social provocation, the intellectual challenge, the freedom for imagination. These groups see only some feared and frightening tomorrow; they differ ever so much from good groups in which children's eyes sparkle with satisfaction.

The age of the child also affects the benefit a youngster will receive from group experience. It sounds so clear-cut to speak of "three year olds" or "four year olds," but this is misleading, of course. All three year olds are not alike, nor is any one child ever "three" in all areas of his functioning. Ideally admission to a group and the length of a child's stay would be decided on an individual basis. Only the need for brevity makes one talk in terms of chronological age alone. The resultant danger of oversimplification must be recognized. However, within this framework it seems accurate to say that the usual three, four, or five year old will gain by being in a good group.

REFERENCES

Aldrich, A. A. and Aldrich, M. M. 1954. *Babies Are Human Beings*. New York: Macmillan.

Axline, Virginia M. 1964. *Dibs: In Search of Self*. Boston: Houghton Mifflin.

Chess, Stella. 1965. *Your Child Is a Person*. New York: Viking.

Dittman, Laura L., ed. 1970. *What We Can Learn from Infants*. Washington, D. C.: National Association for the Education of Young Children.

Fraiberg, Selma. 1959. *The Magic Years*. New York: Scribner's.

Frank, Lawrence K. *Fundamental Needs of the Child*. New York: National Association for Mental Health.

——. 1966. *On the Importance of Infancy*. New York: Random House.

Hymes, James L., Jr. 1963. *The Child under Six*. Englewood Cliffs, N. J.: Prentice-Hall.

——. 1968. *Teaching the Child under Six*. Columbus: Charles E. Merrill.

——. 1971. *Why School before Six?* Arlington, Va.: Childhood Resources, Inc.

Keister, Elizabeth. 1970. *The "Good Life" for Infants and Toddlers*. Washington, D.C.: National Association for the Education of Young Children.

LeShan, Eda J. 1968. *Conspiracy against Childhood*. New York: Atheneum.

Murphy, Lois B. 1962. *The Widening World of Childhood*. New York: Basic Books.

Redl, Fritz. 1966. *When We Deal with Children*. Glencoe: Free Press.

Schulman, A. S. 1967. *Absorbed in Living, Children Learn*. Washington, D.C.: National Association for the Education of Young Children.

Stone, L. J. and Church, Joseph. 1968. *Childhood and Adolescence*. New York: Random House.

Woodcock, Louise. 1952. *Life and Ways of the Two Year Old*. New York: Basic Books.

3

INFANT DAY CARE—
THE OUTCAST GAINS
RESPECTABILITY

BETTYE M. CALDWELL

The Children's Center, established in Syracuse, New York, by J. B. Richmond and myself in 1964, violated one of the then sacred precepts about good day care.[1] It purported to provide day care for infants and toddlers rather than for children in the traditional age range of three to six. Perhaps it was inevitable that the developers would try to rethink the day-care concept, since they were so frequently challenged as to the wisdom of their effort. When forced to be constantly on the defensive, one thinks hard!

The Syracuse Project

The launching of the Syracuse project represented an effective blending of an intellectual idea and a social need. The idea was that the need of a child younger than three for certain growth-fostering characteristics was acute and that society could not afford to postpone the provision of these characteristics. The social need stemmed from the paucity of good child-care facilities for working mothers with children younger than three. But in retrospect it appears that our vision was limited by the constraints implicit in the day-care concept.

COMPONENTS OF THE DAY-CARE PLAN

The following brief quotation is taken from the proposal submitted to the Children's Bureau in 1964 in application for support for the original year of operation of the Syracuse program:

The purpose of this project is to develop and evaluate the effectiveness of a demonstration Day Care Center for young children between the approximate ages of six months and three years. It is anticipated that most of the children will come from low-income homes in which the mother is employed. The basic hypothesis to be tested by this demonstration unit is that an appropriate environment can be created which can offset any developmental detriment associated with maternal separation and possibly add a degree of environmental enrichment frequently not available in families of limited social, economic, and cultural resources.

This summary statement capsulizes our intentions and our conservative yet optimistic hopes. The quotation can serve as a point of reference to trace the evolution of our concept of day care.

A DAY-CARE CENTER FOR CHILDREN UNDER THREE

Day care for children under three was not new, but group day care for this age group was. In states with legislation on the books to regulate day care, facilities offering care to children younger than three generally could not be licensed. The attempt to initiate such a program represented the most unique aspect of the project and commanded the greatest amount of local and national attention. It also aroused the most vehement expressions of concern and opposition, for outstanding leaders in the child-welfare field, who represented the best day care of the period, had gone on record as being opposed to group day care for children under three. In its pamphlet outlining standards for day-care service, the Child Welfare League of America took the firm stand that communities must help parents realize that the very young child should be cared for in his own home by his own mother and should be willing to back this up with financial assistance, homemaker assistance, and counseling if necessary. "If children under three must be cared for outside their own home, they should have individual care in a family and should not be in groups or group facilities." [2]

The original facility developed for this project was small and conservative; we planned for twenty children but with time and space and energy stretched to accommodate twenty-five. Once funding was obtained, the next few steps were easy. This ease of development came about not because of expertise and skill on the part of the staff nor because of any anticipatory wisdom of the early organizers. Rather it was easy to begin because so many parents seemed to need and want such a facility; before our doors were even opened, more than fifty parents indicated their willingness to trust us with their young children. We were fortunate in having demonstration grant funding for the program, as we might never have gotten off the ground if we had to go through regular community planning channels. Almost before we were under way, we were serving a "demonstration" function, for better or worse. Scores of visitors began to visit the program, and this parade never slowed down during the next five years. Undoubtedly the early visitors all saw what they came expecting to see; those who expected to find institutionlike babies raised questions about emotional relationships and asked for reprints apologetically, and those who saw in such a program "the" answer to the educational needs of deprived children went away with an exultant "I knew it could be done," and "Keep me on your mailing list."

CHILDREN FROM LOW-INCOME HOMES

Although day care can be utilized by families from all income levels, there is little doubt that the most acute need tends to be found in low-income families. Not only do these families most need day-care service, but also they are likely to be less able to provide the child with the kind of experiences believed to be essential for development to proceed normally. This assumption was based on a host of descriptive studies documenting childrearing differences in low- and middle-income homes.[3]

The stipulation of our center that the children should have working mothers was a protective one. Although filled with zeal about what the program could conceivably accomplish, we were sufficiently cautious to stipulate that no child would be accepted into the program who was not already in some sort of substitute care. That is, we did not feel secure enough (as no one should before an experiment is done) to advocate group day care for young infants prior to an opportunity to prove that this could be done without undesirable consequences. On the basis of what we knew about the haphazard patterns of substitute care available to low-income families in the Syracuse area, there was little doubt that we could improve on the type of care the children would otherwise have received. Therefore, children of working mothers comprised our target group.

In our original proposal we hypothesized that "an appropriate environment can be created which can offset any developmental detriment associated with maternal separation. . . ." A rereading of phrases like this one forced an embarrassed realization of just how conservative what we imagined to be a radical, even revolutionary, program actually was! The statement is not an assertive boast, but rather a weak and somewhat defensive challenge to some of the overgeneralizations made about the consequences of any type of variance from the sustained dyadic mother-child model. But in fairness the statement represents a logical position in view of the finding that detriment did indeed occur with *prolonged maternal deprivation* and that, if any variation of experience were to weaken the mother-child contact, some tactic to prevent detriment would be necessary. However, at the time the center began operation reinterpretations of the maternal deprivation data had already appeared in the literature and had caused many people to reconsider their stands on the issue.[4] Thus we felt that adequate safeguards for the welfare of the children could be built into the setting. This was going to be accomplished largely by extending the maternal model into the group setting:

Individual attention from adult staff members will be arranged as much as possible for the children in this (younger) group. Continuity and stability of staff will be rigorously maintained. For at least one-half hour in the morning and in the afternoon, each child will receive the concentrated individual attention of one of the full-time staff members. . . . Insofar as is possible, their care upon awakening will be provided by the same staff person.[5]

It was felt that this type of caretaker-child relationship would help prevent the development of anything approximating the maternal deprivation syndrome.

Our original statement also expressed the hope that we could "possibly add a degree of environmental enrichment." This, the understatement of the entire proposal, reflects the conservatism of the prevailing attitudes about group care. We felt that our major commitment was to ensure that the program would not harm the infants and toddlers enrolled in it. Only tentatively did we dare add the hope that the program would possibly add a degree of environmental enrichment! And yet within a short time we were to be branded a "cognitive enrichment" group caring little for social and emotional components of change.

Our proposal for systematic evaluation of the effects of our project was unique for the point in history when the program was launched, for over the years evaluation of the effectiveness of day care has been a pretty casual business. In its pamphlet on day-care standards the Child Welfare League has paragraphs devoted to evaluation in three different sections. Probably the most comprehensive statement is the following:

There should be a continuous process of planning and evaluation with the parents to determine whether the service is of benefit to the child, whether it strengthens and maintains the parents' ability to meet his needs, whether the child or family may need help of some kind, or whether another plan may be more suitable.

Social worker and teacher should have regular conferences, formal and informal, so that they may exchange observations and information about children and families and determine the general direction in which they will be working with individual situations.

Through her own observations, conferences with teachers and others involved in the situation, discussions with the parents, and regular contacts with other community agencies serving the family, the social worker should learn what is happening to the child at home and in day care, to help parents and teachers or the family day-care mother work as harmoniously as possible for his best interests.

Regular conferences of teacher and social worker should be arranged with the staff nurse, and from time to time with the physician. Joint decisions about the appropriateness of home visits in relation to health problems, interpretation of medical recommendations, referrals to clinics, and parent education are among the matters that may be discussed.[6]

This suggests that the effects and the effectiveness of day care should be evaluated by an ongoing clinical process—which, of course, is both valid and valuable. But such a process can hardly provide objective data that will stand up in court or in Congress and demonstrate what happens when children are in day care and which of the changes can be attributed unequivocally to the day-care experience.

This avoidance of systematic evaluation of programs relating to young children has somehow had the approval of most leaders in early child care in this country. In the spring of 1965, when plans for the national Head Start program were first unveiled, the people who had been operating day-care programs for newly labeled disadvantaged children for twenty years should have been able to speak up and temper the optimistic predictions that were being made about how much change could be expected from an eight-week summer program. At that time we were all too excited about the potential value of the proposed program to suggest that perhaps too much was expected. But, as stated in an earlier paper,[7] the fact that we really did not know was "nothing short of a professional disgrace."

We anticipated difficulties in evaluating the impact of the day care to be offered in our project, particularly because the subjects were so young. From the standpoint of the children there were few techniques available that spanned the full developmental period to which our efforts would be directed. From the standpoint of the parents there were few existing techniques that had been standardized on persons with low education and limited reading ability. Fortunately we were simultaneously engaged in ongoing research in early learning, and in that work a large proportion of research effort was devoted to instrument development. Thus in implementing this aim, we had something of an advantage.

As discussed previously, the hypothesis that "maximum advantage may be associated with early stimulation" states the core idea of the whole project, although in discussions of that period we tended to underplay it. Considerable research evidence has converged to suggest that the first three years of a child's life represent the most important period for primary cognitive, social, and emotional development and that it is during this period that the environment will exert maximum effect for either facilitation or inhibition of the child's genetic potential for development. In his book that appeared shortly after our own provisional statement of this hypothesis, B. S. Bloom both formalized the hypothesis and buttressed it with empirical data culled from the psychological and educational literature of this century. Bloom's statement of the hypothesis went beyond our mere prediction of time of maximum sensitivity to an analysis of why the early period was important: "Variations in the environment have greatest quantitative effect on a characteristic at its most rapid period of change and least effect on the characteristic during the least rapid period of change."[8]

There was a valid reason for the cautious statement of the hypothesis in our original proposal, and one that was perhaps unique to the model we had chosen for trying to create an optimizing environment—a *group* day-care program. That is, most of the other persons who were talking at that time about "changing the environment of the young child" were talking in generalities and had faced in only a limited way the question of *how* such changes were to be accomplished. While engaged in this historical narrative, it should be mentioned that we moved to the day-care model only after we had been discouraged with the possibilities of home tutoring and parent education. A review of O. G. Brim's book, *Education for Child Rearing*, which summarized the literature on parent education, imbues one with pessimism about what such programs can accomplish. We moved to the day-care model because to make possible a meaningful test of the effects of additional stimulation, it appeared mandatory that the investigators actually be able to regulate amount and type of stimulation. If parents could not be depended upon to do this at home, then we had no choice but to set up a program in which the care of the children was under our own supervision. Actually we would have preferred that this not be for such large chunks of time as is the case in full day care. But at that time the idea of an *infant* nursery school was totally unacceptable. It was only the increasing pressure from mothers in all parts of the country for better day care for children younger than three that gave the idea a carapace in which to hide when the criticisms began to be hurled.

But this forced a softening of the prediction that "maximum advantage will be associated with early stimulation"—at least to the extent of substituting "may" for "will." For within the day-care model it appeared that we would be dealing with possibly opposing influences on early development. If maternal separation—of the daily, predictable variety with reunion every night—were harmful, with effects presumably less intense than those observed following massive, prolonged separation or deprivation, then any positive gains associated with the environmental supplementation would be reduced by the amount of decrement associated with the separation. It is conceivable that the measured gain in some function might be substantial, whereas the net gain for the child, which takes into consideration certain losses occurring in the new environment, could be minor. Thus any prediction about overall change should perhaps be stated conservatively.

An attempt was made to deal with these opposing influences for change by limiting the program to children who were at least six months old at the time they enrolled. This policy decision was made on the basis of data available from studies of infant attachment, which suggested that by this age the infant's attachment to his own mother is fairly well developed.[9] The reasoning was that, once this attachment had developed, placing the child in a situation where he would be exposed to many adults in the course of a day would not distort his basic emotional attachment to his family. Some time later Anna Freud challenged the logic of this policy, suggesting that the point in time at which an attachment had just been developed was not a propitious time to place the attachment in a vulnerable situation.[10] This point is worth mentioning merely as one other reason for the essentially conservative nature of our prediction about the inverse relationship between the age of beginning environmental enrichment and the extent of response to the enrichment.

PARENT INVOLVEMENT

The initial proposal said nothing about parent involvement. However, our first publication at least paid a little attention to the idea:

> An active parent education program will be a regular part of the activities of the Center. . . . Monthly conferences between parents and the Director of the Center or individual teachers will be scheduled, at which time the child's development can be discussed with the parents. More intensive casework will be offered those families needing such service. Each parent will be asked to "volunteer" a few hours a week in the Center. This will give the staff an opportunity to observe the mother's pattern of interaction with her own infant and with other infants, and will also give each mother a chance to observe the child care practices of the Center staff members.[11]

While that does not exactly say that we would ignore the parents, it seems in retrospect that the attitude is there. (And here I wish to make it clear that any naïveté in the statement should be interpreted as representing mine and not that of my colleague of that period, Dr. J. B. Richmond.) In defense of this narrow concept of parental involvement, I can only suggest that it contained the germ of the enlarged concept that was soon to develop. For, as will be developed further, in *some* day-care settings with *some* families it is not feasible to develop a symbiotic working relationship with the parents. The home and the day-care center may work at cross-purposes no matter how valiantly the staff might try to make it otherwise.

HISTORICAL SUMMARY

In summary, the decision to use group day care for infants was arrived at somewhat by the conceptual backdoor: the pressures from working mothers with young children for good day care for their infants provided an extremely relevant social context in which some important theoretical ideas about early development could be tested. This timely fusion of intellectual ideas and social needs may well represent a necessary, if not sufficient, condition for significant conceptual advance in the social sciences.

Evolution of an Expanded Day-Care Concept

FRIENDLY VOICES OF SUPPORT

Gradually we came to realize that the implications of the Syracuse program transcended anything that we had been aware of at the time the center was established. What was originally thought of as a socially acceptable way to test a mildly controversial scientific hypothesis was soon perceived to be a potential instrument of major social policy. In short we began to think hard about the implications of what we were doing.

And we were certainly not the only ones. At the National Conference on Day-Care Services held in Washington, D.C. on May 13–15, 1965, Mrs. Katherine B. Oettinger, at that time chief of the Children's Bureau, in a keynote address urged some flexibility in thinking about day care for the very young. To quote her:

> The Children's Bureau and other leaders, in their profound belief in supporting the mother-child relationship, have discouraged any separation in the early years. We have followed inconclusive and groping scientific opinion that, in the past, has advised against mother-child separation in these early child developmental stages. . . .
> But experience and research by leading psychiatrists and child care experts have revised that narrow gauge viewpoint. Now, most careful investigators tell us that it is deprivation or a deficit of mothering, not a separation from the mother, that harms the young child. . . .
> Reality also has forced us to revise our attitude. As often as we have insisted that the well-being of the young child depends on the mother remaining in the home, she has continued in increasing numbers to work outside the home. . . .
> While the problem grew with the expanding number of working mothers, we refused to act because we thought that any solution except keeping the mother in the home or the infant in a substitute home was improper. . . .
> Mostly, mothers have had to find their own makeshift arrangements because our whole philosophy has been to sweep the problem of infant day care under the rug of disapproval.[12]

She then went on to make an announcement about the availability of child-welfare grants, under which the Syracuse project had been funded, and to mention the establishment of the Children's Center as one approach in an attempt to fill the gap in the availability of quality day care for infants.

About this time the Child Welfare League also softened its position somewhat. In a revision of their pamphlet on day care D. B. Boguslawski made the following statement:

> Children under three also need supplemental care for some hours each day. Experience has shown, however, that children so young are generally not ready for group living. They are not mature enough to play with other children: they need a highly personalized individual relationship with an adult and can suffer if they do not have it. Some research-demonstration centers now in operation are providing group care for children of this age; these centers have a high ratio of staff to children and a continuing evaluation by psychologists, psychiatrists, and social workers. It is too soon to know whether the optimum conditions will be found under which infants can be cared for in groups without damage to their personality development. For the present, family day care is considered the best supplementary-care resource for children under three.[13]

These indications that other persons were also rethinking the implications of the day-care concept were unquestionably influential in helping to guide our own nascent conceptual restructuring.

EVIDENCE OF VALIDITY OF THE NEW CONCEPT

Part of our original commitment had been to evaluate the consequences of participation in our day-care program. The original research plan called for follow-up of the children until they were at least seven years of age, a point in time that will not have been reached by all the children for several years. However, we now have in hand ample evidence relating to the effects of the day-care experience in several areas of development. Detailed reports of the analyses of these data are either already completed or in preparation, so only highlights will be presented here.

Maternal Attachment

Maternal attachment was one of the most crucial areas in which an evaluation of the early day-care experience was needed. On the basis of careful interviews with the mothers of twenty-three home-reared and eighteen day-care children, all of whom were approximately thirty months of age at the time of the interview, ratings were made on seven scales comprising an operational definition of attachment. Essentially no significant differences between the groups could be detected in terms of attachment of the infants to their mothers. However, significant associations were found when the home-care and day-care samples were pooled between the strength of attachment (both child for mother and mother for child) and the developmental level of the child and attachment and the amount of stimulation and support for development available in the home.[14]

Emotional Disturbance

Sixteen children enrolled in the four-year-old group at the center were evaluated by a trained child psychiatrist in terms of a five-level scale of personal and social adjustment. Half of the children had enrolled after the age of three, and half had been enrolled for periods ranging from one to three years. Three-fourths of the children were rated as falling within the top three categories representing adequate to excellent adjustment, and the distribution of children from the under-three and the over-three subsamples across the five levels of adjustment was random. That is, there was no association between level of adjustment and age at entry into the day-care program.[15]

Cognitive Changes

Between the fall of 1964 and the spring of 1968 some eighty-six children younger than three at the age of enrollment had remained in the program for at least six months (most of them for more than one year, with an average duration of twenty-five months).[16] At the time of enrollment these children had a mean age of seventeen months and earned an average developmental quotient of 102.7. When retested in the spring of

1968 the average developmental quotient was 116.5, a highly significant increase. These gains were comparable for those shown by a group of twenty-two children older than three at the time of enrollment (average of forty-four months)—from 101.8 to 119.3 after a mean participation interval of 12.6 months. Differences between changes shown by both the children under three and those over three and their control groups were statistically significant beyond the .01 level of confidence.

None of the children from the project are now old enough to enable us to say anything about the durability of these changes. Our initial hypothesis would suggest, for example, that the children who entered day care younger than three but who showed no greater cognitive gains after a mean interval of twenty-five months than the older entries did after a mean interval of 12.6 months would nevertheless be able to sustain those gains longer and be better insulated against future environmental inadequacies. Proof of that hypothesis must await the further chronological development of the children, however. For the present all we can claim to have demonstrated is that the younger children are certainly not cognitively damaged by the day-care experience and that they are indeed benefited by it.

FROM CARE AND PROTECTION TO ENVIRONMENTAL SUPPORT

Although we were ostensibly testing the ability of the environment to counteract forces associated, with developmental decline, our "charter" as a day-care center specified that our task was to provide care and protection for the children participating in the program. Referring once again to the widely used guidelines for evaluating day care, we find the statement, "The primary purpose of the day-care service is the *care and protection* of children." [17] Perhaps no one would challenge this statement, but care and protection mean different things depending upon the breadth of the social context in which they are placed. At the time the day-care concept gained adherents and momentum, the types of social hazards from which we wanted to protect young children were things like insufficient food, lack of shelter, or physical abuse. But with today's knowledge about developmental influences, care and protection must also include arranging for the child to have the essentials of experience. By focusing too much on the avoidance of developmental negatives, the early day-care movement may well have truncated its potential social influence.

Gradually we began to realize that staff relationships with the children could potentially have even more impact than we originally visualized. We had these children in our presence from early morning until early evening; often they went directly to sleep shortly after they left our "care and protection." What they got from us was to a great extent what they got from that day. Thus instead of thinking of ourselves as a "day-care" facility, we began to see our program as representing a supplementary developmental environment containing an appropriate

amount and variety of growth-fostering experiences. Instead of being
content to remain a "center" where we "cared" during the day, we
wanted to fulfill the opportunity to create a normalizing, or optimizing,
developmental supportive environment. The implication for staff activity
was that we needed to devote a major portion of our time and energy to
the development of a program model. To offer day care is one thing; to
be able to specify how the day is to be spent, how to "govern the encounters that children have with their environments" is quite another.[18]

REDEFINING THE POPULATION TO BE SERVED

Just as the concept of day care under which we were functioning
came to be constraining, so did our own specifications about the population we should serve. Accordingly we have modified our admission criteria in terms of age, social class, employment status of the mother, and
number of hours per day spent in attendance.

Extending the Age Range

After only one year of operation it became painfully obvious that
there were not enough standard day-care facilities in Syracuse to absorb
the first year "graduates." Also, after the first year of operation some of
our earlier naïveté about programing had vanished, forcing us to realize
that the children could not simply be transferred into any program and
receive the same quality of care. In order to be able to answer our question about the effectiveness of environmental supplementation as a function of age of beginning the supplementation, it would be necessary to
have children entering the same type (though, of course, regulated to
developmental level) of program at all points during the early childhood
years. Therefore, we began to revise our arrangements to permit an extension of the program to cover three and four year olds. This was made
possible in the fall of 1966 by receipt of an enlarged grant from the
Children's Bureau. But even this extension was not enough, as working
mothers with a child of kindergarten age cannot manage with only the
short half-day of care provided by most public-school kindergartens. (Of
course, in states where there are no public kindergartens, most day-care
centers include five year olds in their regular program.) We are now
moving to the point of realizing that, if the day-care program truly offers enrichment rather than merely care and protection, some degree of
contact even after the child is in primary school is essential.

Broadening the Social Class Range

Our intent to seek to circumvent developmental decline in children
for whom such decline could be predicted defined our sample as consisting primarily of children from lower social class backgrounds. No sooner
were we functional, however, than we began to question the wisdom of
this policy decision.

Soon after we opened we were pressured by other community agencies and by parents themselves to accept children from nondeprived families. We received many referrals from "marginal" middle-class families

(mother in mental hospital, father with full child-care responsibility; college student deserted by young husband, living with year-old baby in abject poverty; postal worker's family with so many children mother showed "worn out" syndrome) whose needs were every bit as acute as those of more culturally deprived families. Also, we received a fairly large number of applications from middle-class parents who liked the idea of such a facility and simply wanted their children to enjoy the advantages of it. Some in fact demanded "equal numbers."

It was difficult to turn a deaf ear to such applicants. In addition we began to question the wisdom of having a facility that would represent social, if not racial, segregation. Furthermore, we found ourselves using words like "pacers" and wondering whether we needed a few bright, advanced, and highly verbal children in every group to serve as stimuli to the other children. Therefore, it became our official policy in the fall of 1966 to include approximately one-third middle-class children in our groups. In general, preference was given to middle-class families with problems, but in a few instances our new enrollees came from backgrounds as free of visible pathology as could be imagined.

At this juncture we are unwilling to propose any social class limitations except as these can be determined on the basis of greater or lesser accessibility to alternative resources. Middle-class children respond beautifully to our supplementary environment. In fact most of them literally soar in it. In an earlier summarization of the literature on the effects of preschool experience, J. W. Swift concluded that, by and large, the greater the disadvantage in the home environment, the greater the gains associated with preschool.[19] In our data, especially for the three- and four-year-old groups, the pattern tends in the opposite direction—both groups show positive gains, but the middle-class gains are even more impressive. This finding has major implications for planning supplementary environmental experiences for young children from all types of family and social backgrounds. Most of our early intervention projects are based upon what F. D. Horowitz and L. Y. Paden have called the deficit model [20]—we assume that certain deficits exist for which we can compensate. Likewise, we tend to assume that the middle-class child will not have these same deficits and that his development, as determined by observed behavior or test performance, will reflect an output that is closer to the maximum of which the child is capable. The fact that this is apparently not the case has important implications for the planning of educational facilities for all young children.

Including Children of Nonworking Mothers

After our first year of operation, following inspection of data relating to patterns of change associated with enrollment in the program, we became sufficiently secure to drop the requirement that mothers of children in the center had to be fully employed. At first any nonworking mother whose child was accepted had to represent extreme social and cultural deprivation. However, we soon relaxed the requirement about maternal employment even more. For the most part children of non-working mothers enrolled in the center at any one time tended to represent mothers who had worked at the time the child was initially enrolled but who had either quit or lost their jobs. In the course of a full intake procedure (detailed social history, medical examination, psychological examination, home visit, observation of mother-child interaction) we devoted many hours to a given family. Once a child was in attendance the time investment climbed rapidly. In view of the job instability of many women in low-income occupations, rigid adherence to the maternal employment criterion could have meant a constantly changing sample, with no one remaining in the center long enough to evaluate the changes associated with attendance. Therefore, we made the policy decision to keep any child whose mother had no objection to this arrangement. In four years time not a single mother has objected. At this juncture we are willing to go even further and make the service available to some children whose mothers have never been employed and may never be employable.

Varying the Length of the Day

Once intake policies had been changed to permit inclusion in the sample of children of nonworking mothers and children of middle-class mothers, we were not bound by the stricture of full-day care. As young children tend to sleep during much of the afternoon, it seemed possible that the children gained less from contact with the program during the afternoon than during the morning. Therefore, whenever a child's mother did not work, we often made the arrangement to have the child attend the center only in the morning (usually remaining through the lunch hour) and return to his home for the afternoon. With this change we were closer to offering "day care by the yard," as suggested by one visitor. When enough children have been through the program to permit analysis by age collapsing across cohorts, it should be possible to determine more objectively the consequences of such variations.

Designing a Supportive Environment

It did not take long to reassure us that the emotional dangers associated with infant day care had been, like reports of Mark Twain's death, greatly exaggerated. Daily separation followed by nightly reunion appeared to have none of the same effects as total separation over a period of weeks or months. At a Conference on Group Care for Infants in 1967, Anna Freud acted astounded that we would have even bothered to worry about it. She reminded us that there was no comparison between the situation where a child has no parents (as in an institution) and one where he has parents but spends part of each day away from them. Such opinions, plus our own data,[21] helped us to relax our somewhat defensive posture, and we moved consciously toward an attempt to design a supplementary environment conducive to optimal development. Our new concept of care and protection found expression in the phrase "the supportive environment." [22]

This period was characterized by functional ambivalence about our own objectives, regardless of the bold statements we were putting into print. We continued to seem uncertain about whether we must show the child-welfare field that children participating in our program were not harmed, or whether we could hope to demonstrate that they showed significant developmental gains. During this period I wrote two papers that represented an attempt to synthesize my own thoughts about environments conducive to development. The titles are worth mentioning, as they clearly reflect the switchover from negative to positive thinking. The first was entitled, "What Is the Optimal Learning Environment for the Young Child?" and simply raised the question of whether the optimal environment for a young child was necessarily the constant company of his mother if the relationship provided insufficient emotional support and lacked the extent and variety of sensory input necessary to assist the child's cognitive growth.[23] The second was entitled "On Designing Supplementary Environments for Early Child Development" and attempted to formulate principles that represent something of a least common denominator of conditions necessary for healthy development.[24] The thought was that such a set of general principles would be helpful in the structuring of the total supplementary environment in which early enrichment experiences were offered, regardless of the specific teaching curriculum (for example, Montessori, Bereiter-Engelmann, and so forth) that might be followed. Stripped of empirical supports, these principles are as follows:

1. The optimal development of a young child requires an environment ensuring gratification of all basic physical needs and careful provisions for health and safety.
2. The development of a young child is fostered by a relatively high frequency of adult contact involving a relatively small number of adults.
3. The development of a young child is fostered by a positive emotional climate in which the child learns to trust others and himself.
4. The development of a young child is fostered by an optimal level of need gratification.
5. The development of a young child is fostered by the provision of varied and patterned sensory input in an intensity range that does not overload the child's capacity to receive, classify, and respond.
6. The development of a young child is fostered by people who respond physically, verbally, and emotionally with sufficient consistency and clarity to provide cues on appropriate and valued behaviors and to reinforce such behaviors when they occur.
7. The development of a young child is fostered by an environment containing minimal restrictions on exploratory and motor behavior.
8. The development of a young child is fostered by careful organization of the physical and temporal environment that permits expectancies of objects and events to be confirmed or revised.
9. The development of a young child is fostered by the provision of rich and varied cultural experiences rendered interpretable by consistent persons with whom the experiences are shared.
10. The development of a young child is fostered by the availability of play

materials that facilitate the coordination of sensorimotor processes and a play environment permitting their utilization.

11. The development of a young child is fostered by contact with adults who value achievement and who attempt to generate in the child secondary motivational systems related to achievement.

12. The development of a young child is fostered by the cumulative programing of experiences that provide an appropriate match for the child's current level of cognitive, social, and emotional organization.

Recently other attempts to formulate characteristics of growth-fostering environments have been published,[25] varying from this formulation in simplicity, elegance, or formal theoretical sophistication. Yet there are enough consistencies in the different formulations to persuade one that it is possible to be reasonably specific about the type of environment we wish to design.

But to return to the main theme of this chapter, would it be possible to provide all of those features in a program of infant or preschool day care? Is it possible to concretize such principles in our down-to-earth, poorly named, institutionalized social setting known as day care? The answer to my question would have to be "yes," provided there is some degree of continuity between the home environment and the day-care environment.

Expansive Day Care

The voices of criticism that sounded forte during the early days of the Syracuse day-care program are now heard infrequently and very pianissimo. In fact the main criticism now voiced is that day care for the very young is very expensive. Too expensive it is; too expansive it is not.

Whether we like it or not, society *is* intruding ever more into the domain once considered the exclusive province of the family. To use Charles Gershenson's term, it is coming to serve the function of a third parent.[26] We need to give careful thought to how it will play that role.

As more and more programs for infants and young children materialize, we must give more thoughtful attention to the full implications of such programs. It is not enough just to talk about "curriculum design," "fostering mental health," "testing the Piaget model," "using behavior modification," or "developing a primary prevention model." In such programs we are acting in loco parentis, and we need to make certain that we are wise parents. None of the above principles of growth-fostering environments can mean anything unless they are applied in an environment in which the training adults can be explicit about what their goals for childrearing are. Do they want obedient children? Happy children? Militant children? Bright children? Group-oriented children? Individuals? It is foolish to talk about designing an optimizing environment unless we face up to the chore of trying to specify some of the proximal and distal goals of our child training. In our culture we seem to be victims of the implicit assumption that this task will take care of itself.

Many people who visited the Syracuse program remarked to me, "It's like a little kibbutz, isn't it?" And although I cannot answer with the force that might be possible had I actually visited a kibbutz, day care in America comes closer to the kibbutz concept than anything else we have. But American day care is still somewhat paralyzed by its conservative past—to protect the welfare of young children. It has yet to face up to the challenge of what it can become—the best existing resource for attempting to design and actualize environments capable of fostering growth during the very early years of life.

NOTES

1. The project described in this chapter was supported from its inception by Child Welfare Research and Demonstration Grant No. D-156 from the Children's Bureau, U.S. Department of Health, Education, and Welfare. The author's current work is supported by Grant No. SF-500 from the same source.

2. Child Welfare League of America, Standards for Day Care Service. (New York: Child Welfare League of America, 1960), p. 6.

3. See H. Wortis, J. L. Bardach, R. Cutler, A. Freedman, and A. Rue, "Child Rearing Practices in a Low Socioeconomic Group," Pediatrics 32 (1963):298–307; E. Pavenstedt, "A Comparison of the Child-Rearing Environment of Upper-Lower and Very Low-Lower Class Families," American Journal of Orthopsychiatry 35 (1965):89–98; C. A. Malone, "Safety First: Comments on the Influence of External Danger in the Lives of Children of Disorganized Families," American Journal of Orthopsychiatry 36 (1966):3–12; C. S. Chilman, Growing up Poor (Washington, D.C.: U.S. Department of Health, Education, and Welfare, Welfare Administration, 1966).

4. L. J. Yarrow, "Maternal Deprivation: Toward an Empirical and Conceptual Re-evaluation," Psychological Bulletin 58 (1961):459–490; L. J. Yarrow, "Separation from Parents during Early Childhood," in M. L. Hoffman and L. W. Hoffman, eds., Review of Child Development Research (New York: Russell Sage Foundation, 1964), 1:89–130; M. D. S. Ainsworth, "Reversible and Irreversible Effects of Maternal Deprivation on Intellectual Development" (New York: Child Welfare League of America, 1962), pp. 42–62.

5. B. M. Caldwell and J. B. Richmond, "Programmed Day Care for the Very Young Child—A Preliminary Report," Journal of Marriage and the Family 26 (1964):481–488.

6. Child Welfare League, op. cit., p. 17.

7. B. M. Caldwell, "On Reformulating the Concept of Early Childhood Education: Some Whys Needing Wherefores," Young Children 22 (1967):348–356.

8. B. S. Bloom, Stability and Change in Human Characteristics (New York: John Wiley and Sons, 1964), p. vii.

9. H. R. Schaffer and P. E. Emerson, "The Development of Social Attachments in Infancy," Monographs of the Society for Research in Child Development 29 (1964):1–77; J. Bowlby, Attachment and Loss (New York: Basic Books, 1969).

10. A. Freud, "Comments on First Day's Reports," in H. Witmer, ed., On Rearing Infants and Young Children in Institutions, Children's Bureau Research Reports, no. 1 (1967), pp. 47–55.

11. Caldwell and Richmond, op. cit., p. 487.

12. K. B. Oettinger, "A Spectrum of Services for Children," in *Spotlight on Day Care*, Proceedings of the National Conference on Day Care Services, Children's Bureau Publication, no. 438 (1966), pp. 121–134.

13. D. B. Boguslawski, *Guide for Establishing and Operating Day Care Centers for Young Children* (New York: Child Welfare League of America, 1966), *p. J*–52.

14. B. M. Caldwell, C. M. Wright, A. Honig, and J. Tannenbaum, "Infant Day Care and Attachment," *American Journal of Orthopsychiatry* 40 (1970):51–52.

15. S. J. Braun and B. M. Caldwell, "Social Adjustment of Children in Day Care Who Enrolled Prior to or after the Age of Three," in preparation.

16. An analysis of cognitive changes associated with participation in the program is currently being made for all children enrolled between 1964 and 1969. That is, we are collapsing dates of entry into the center and examining effects that can be measured at some specific evaluation point for children entering the day-care program at different ages. But partial evaluations have been made repeatedly in order that we might have immediate feedback about effects associated with participation in the program.

17. Child Welfare League, *op. cit.*, p. 2.

18. J. Mc V. Hunt, *Intelligence and Experience* (New York: Ronald Press, 1961).

19. J. W. Swift, "Effects of Early Group Experience: The Nursery School and Day Nursery," in Hoffman and Hoffman, *op. cit.*

20. F. D. Horowitz and L. Y. Paden, "The Effectiveness of Environmental Intervention Programs," *Review of Child Development Research*, (New York: Russell Sage, 1970) vol. 3.

21. Caldwell, Wright, Honig, and Tannenbaum, *op. cit.*

22. B. M. Caldwell, "The Supportive Environment Model of Early Enrichment," paper presented at the meeting of the American Psychological Association, Washington, D.C., September 1969.

23. B. M. Caldwell, "What Is the Optimal Learning Environment for the Young Child?" *American Journal of Orthopsychiatry* 37 (1967):8–20.

24. B. M. Caldwell, "On Designing Supplementary Environments for Early Child Development," *BAEYC Reports* 10, no. 1 (1968):1–11.

25. S. Provence, *Guide for the Care of Infants in Groups* (New York: Child Welfare League of America, 1967); S. Provence, "The First Year of Life: The Infant," in L. L. Dittmann, ed., *Early Child Care: The New Perspectives* (New York: Atherton Press, 1968), pp. 27–39; L. B. Murphy, "Individualization of Child Care and Its Relation to Environment," in Dittmann, *op. cit.*, pp. 68–104; J. L. Gewirtz, "On Designing the Functional Environment of the Child to Facilitate Behavioral Development," in Dittmann, *op. cit.*, pp. 169–213.

26. Personal communication.

PART II

---

*Child Care in
the United States*

4

FROM A BLACK PERSPECTIVE:
OPTIMUM CONDITIONS FOR
MINORITY INVOLVEMENT IN
QUALITY CHILD-DEVELOPMENT
PROGRAMING

THE BLACK CHILD
DEVELOPMENT INSTITUTE

The Black Child Development Institute (BCDI)[1] is guided in its analysis of legislation by an organizational commitment to building child-development institutions that meet the comprehensive needs of Black children, families, and communities.

Our premises are threefold. First, child development must encompass not only the Black child's cognitive, physical, and social needs, but his psychological requirements as well. Black and white scholars alike have focused upon the need to instill confidence, self-knowledge, and self-respect in the very young to motivate them for educational achievement and individual accomplishment. Therefore, there is a need for curriculum and program content that will enhance the Black child's sense of racial awareness and pride and will reflect the many contributions of his race to the culture and history of the American Republic.

Second, Black people and other minority groups are increasingly aware of the failure of the American educational system, controlled by whites, to prepare minority children for full participation in the society at large. In the face of this institutional failure, Black people have begun to insist that the decision-making authority to develop educational policy tailored to their children's distinctive needs must rest with the community affected. The desire for community control of day-care and child-development programs is born of the realization that the first five years of life are too critical to be entrusted to those for whom the Black child's interests may not be predominant.

Third, child-development centers can be catalysts for *total* community development. Community controlled institutions have the potential to facilitate the *multiple* use of institutional resources for economic and social growth. The exercise of such power and resources on behalf of group self-interest has been the historical means by which other minority groups have achieved individual and collective progress.

The following series of proposals flow then from the assumption that there is a need for legislative and administrative policies and regulations that will maximize the involvement of Black and other minority communities in all aspects of child-development programing. So that program efforts reflect .the Black perspective, it is essential that minority groups be given an opportunity to make basic policy decisions at all levels, in communities where services are provided as well as in government agencies where funds are distributed. Program quality will result, for the Black Child Development Institute is convinced that consumer control can be a powerful force for excellence. Community empowerment can provide effective leverage for the improvement of the overall social and economic condition of the nation, as it becomes a means of ensuring the equitable participation of minority groups in American society.

What Are Child-Development Programs?

INDIVIDUAL AND COMMUNITY DEVELOPMENT

Essential to the creation of meaningful child-development programs is an accurate definition of such programs. "Developmental" programs make possible both the optimum comprehensive development of the child as well as the optimum economic and social advancement of his family and the specific community from which the child comes. Such programs should address the total needs of young children—physical, social, intellectual, and emotional—but only as determined by the parents and leadership of the community. In other words the efforts and achievements of a child-development center are complementary, not interventionary. A child-development center should assist in *strengthening* the community-parent-child relationship, but must not interpose itself as a community-parent substitute in that relationship. In addition child-development projects can have great economic and social impact if they are implemented as *community* institutions that seek every opportunity to catalyze community development and change. Thus project staffing, construction, remodeling and renovation of facilities, training, the provision of services, supplies and equipment, and the participation and training of parents are not isolated activities, but are opportunities to maximize the center's social and economic impact upon the community and its members.

VARIETY OF FUNDABLE PROGRAMS

A related consideration is the type of program that should be eligible for federal assistance. Using the definition above, a wide range of programs could be funded, ranging from those that maintain close parent-child ties (mini-programs serving children of a given family or extended family) to programs that serve a citywide constituency. Special considerations might be given to existing programs (for example, the neighborhood baby-sitter who cares for her own and her neighbor's children), with funds and technical assistance provided for the upgrading of service and maintenance of quality standards.

Programs should also be scheduled flexibly to meet the needs of the communities in which they are located, whether they are needed for part of the day (after school) or on a twenty-four-hour basis. Ages of eligible children may vary with the kind of program available; ideally funds should be made available to meet the needs of youth under sixteen. Federal standards should be flexible enough to permit the funding of specialized programs to meet the needs of infants as well as older children, providing that adequate safeguards and guidelines are observed.

ELIGIBLE SPONSORS

In the field of child development new institutions with fresh and innovative approaches are badly needed to meet existing demands. Eligible sponsors of child-development programs should be private and public, nonprofit and profit-making organizations and corporations. Incentives should be given to applicant agencies, organizations, corporations, or cooperatives comprised of those to be served by government child-development programs or their representatives. They should be encouraged and assisted in competing with established public and private agencies.

Appropriation-Authorization

ADEQUACY

Ideally projections of costs for child-development services are based upon the fact that there are over three million economically disadvantaged children who are presently not being served by day-care and child-development programs. In addition only 2 percent of working mothers have access to group day-care facilities, and there are over five million children under five whose mothers work during the day.

Estimated costs per child per year run from an average of $1,196 (based on average projected costs of twenty-three dollars weekly as calculated by franchise industry spokesmen) to desirable costs of $2,320 (projected by government representatives as the minimum needed for comprehensive programs). If new legislation in this area seeks to significantly address the community's real needs, and expansion were to occur at a rate sufficient to reach an enrollment of only 800,000 children the first year of a program's operation, "federal obligation would be $1.1 billion plus whatever portion of $850 million in construction costs would be federally supported." [2]

DISTRIBUTION AND ADMINISTRATION OF AUTHORIZED FUNDS

Community Decision-Making:
The Primary Role of the Parent-Consumer

The primary function of a delivery system must be to guarantee that program objectives and goals are met at the level where services are received. In child-development programs this interaction occurs at the community level. Parents and the advisers they choose must therefore be responsible for policy decisions concerning centers where their children are enrolled. Accordingly the optimum mechanism for the distribution of funds provides for a direct relationship between the federal government or other funding sources and the consumer controlled child-development center that provides the service. If screening of proposals or fund requests must occur (for example, to determine if program standards will be met or to distribute limited funds equitably), such review should be conducted by policy-making committees or councils as close to the parent-consumer as possible, both geographically and representationally. No less than two-thirds of these local policy bodies should be made up of parents, with the remaining one-third to be selected by the majority.

Federal Administration

It follows then that there should be minimal involvement in the distribution and administration of funds by various levels of government. Optimally funds would be distributed by a federal agency or agencies through representatives in regional, state, or local offices, who would distribute funds upon receipt of proposals from project applicants or local policy boards, in accordance with specific program guidelines. Such guidelines should require the funding, at the minimum, in each geographical area of a representative percentage of minority operated or owned organizations, corporations, and cooperatives.

The federal government must provide information and technical assistance *in advance* to community-based minority organizations and groups, so that they can develop proposals and funding plans that meet federal criteria. Recommended would be the use of minority contractors and consultants to perform this task.

Provisions must also be made for the creation of policy and advisory committees at national and regional levels, so that minority groups might have meaningful input at all levels of governmental programing. Parent and minority group representatives should constitute no less than 66⅔ percent of the total membership of such bodies.

Governmental employees charged with administering child-development programs shall be representative of the population to be served by such programs. There must be provisions for the representational participation of minorities in policy-making positions.

The formula for the distribution of operational funds should conform to the patterns and concentrations of need in regions, states, and local communities (for example, allocated on the basis of the percentage of economically disadvantaged children in relation to the total number of children). Certain funds must also be retained by the responsible federal official for purposes of administration, demonstration, direct funding of projects, technical assistance, research, and evaluation.

Decentralized Administration: Municipal Prime Sponsorship

In the event funds must be channeled through an intermediary decision-making apparatus, BCDI supports the concept of an areawide prime sponsorship arrangement at the municipal level if:

1. There is no population limitation so that rural townships as well as large cities might be eligible.
2. Policy-making is entrusted to communitywide councils, two-thirds of which are representatives of neighborhood groups. The balance of these councils would be selected by the two-thirds majority from a list of nominees provided by the prime sponsor.
3. The responsible federal official is permitted to retain 15 percent of the overall appropriation for the direct funding of projects that are not processed by prime sponsors.
4. There are provisions that permit nonprofit organizations and agencies to be given preference as prime sponsors when municipalities have maintained a pattern of exclusion of minority groups or insensitivity to the needs of economically disadvantaged citizens, or when substantial objections are raised by representatives of such groups to the application of such units for prime sponsorship responsibility.

State Involvement

The involvement of states in the administration of federally funded child-development programs *is not desirable*. To be sure, some few states are spending funds for child-development programs equal to or more than the federal investment. Most state bureaucracies, however, are governmental entities particularly remote from the people to be served, unresponsive, and administratively ill-equipped to develop innovative programs. Should state involvement and funding be unavoidable, however, certain safeguards should still be guaranteed. (See section on safeguards in the event of state involvement below.)

REPEAL OF OTHER AUTHORIZATIONS

There is justification for the repeal of other legislative authorizations that permit the expenditure of funds for child care if the agency designated to consolidate those authorities is permitted to preserve the strengths of other programs (for example, parent participation in Head Start) and has the sensitivity and ability to meet community needs more adequately than the separate agencies now do. To centralize for efficiency alone is not a sufficient argument for consolidation.

Once this condition is met, there can be further justification for consolidation and repeal of other authorizations only if the new dollar amounts proposed *clearly exceed* those presently available under separate authorizations. Thus support of existing programs is continued while permitting the funding of new projects as well.

FEDERAL MATCHING FORMULA

Given the minority community perspective, the optimum situation is a federal contribution for project or contract costs of 100 percent of costs incurred. If less than 100 percent is provided (minimum 90 percent), the matching share can be met with in-kind contributions. Federal regulations regarding in-kind should be geared to what community project sponsors can realistically produce.

FEE-SETTING AUTHORITY

The responsible federal official should have authority to set minimum and maximum fee levels where fees are charged. (Children who are economically disadvantaged should *not* be required to pay any fees.) State officials, who are expected to be more vulnerable to entrepreneurial pressures, should *not* have such authority. Moreover, if fees are counted as the nonfederal matching share, the federal official should be required to provide for the frequent monitoring of programs in which noneconomically disadvantaged children are enrolled to assure that the needs of the economically disadvantaged are still being given priority.

Funding Procedures

CONTINUITY OF FUNDING

One of the major sources of frustration and concern in minority communities has been the one-year funding cycle, which has made planning, experimentation, and recruitment of staff difficult. New legislation should permit funding cycles of at least three years, with provisions for the yearly evaluation of programs. This approach is particularly appropriate in view of the institutional character of the child-development center. Programs found wanting in the interim could be revised, given technical assistance where there are shortcomings, or discontinued.

DIRECT PROJECT GRANTS AND CONTRACTS

Eligible applicants should qualify for federal support by submitting proposals for the funding of comprehensive child-development programs. The use of this funding approach permits community-based organizations that do not have start-up capital to provide day-care services. This approach also reinforces the principle of the *right* of children to these services, ideally a right that should be extended to *all* children. However, it is important to provide for maximum parental choice within this system. Parents should not be limited, for example, to the use of a center that serves their geographic area.

If funding approaches in addition to or other than the project grant or contract system are proposed, both the pros and cons of such methods should be assessed. Two major considerations are:

1. What approach maximizes the opportunities for communities to develop day-care centers that meet their collective needs?
2. Which method maximizes the parent's free choice in securing day-care services?

VOUCHER SYSTEM

A voucher system permits the provider of day-care services to be *reimbursed* by the government for the costs of the care, after submitting a voucher given to him by the consumer (for example, food stamps). A mandatory voucher system that requires parents receiving assistance to place their children in approved facilities is unacceptable for these reasons:

1. Community organizations could be handicapped in developing approved facilities to qualify for reimbursement under a voucher plan because of lack of investment capital.
2. The need to certify facilities, usually done by municipalities and states, could permit the intrusion of *subjective* judgments, unless standards for certification were uniform and a monitoring system was built in.
3. Governmental agencies would be interposed, depriving the individual of the widest possible range of options in determining what would be best for his child.

The voucher system that was combined with other systems (for example, direct project grants) and that permits collective action by individuals might provide an acceptable experimental approach. For example, if fifty parents would be permitted to combine their vouchers and thus certify that they intended to develop a new facility or utilize an existing one, such certification could result in a project grant for that group.

CASH PAYMENT

Another approach permits an affected individual (for example, a Family Assistance Plan enrollee) to receive a bonus cash payment to provide services for his children. Although there could be no guarantee that funds would be spent for that purpose, such individuals would have the same free choice as all other citizens. The *exclusive* use of this method would make it difficult, however, for community organizations to develop a facility to qualify for funds under this method since start-up costs could not be guaranteed.

FUNDING METHOD EXPERIMENTATION

If the use of funding methods (other than exclusive reliance on project grants and contracts) is legislated, all methods should be tried in a given community on a carefully controlled basis subject to federal approval and review. Organizations in municipal districts of certain sizes (for example, 5,000 families) could qualify to receive project grants. In addition individuals in that community might be given a choice of using the facilities funded by project grants or the voucher or cash payment system. In smaller municipalities fewer options might be exercised. None of the methods other than the grant or contract approach should be mandated, however, until they are tried and evaluated.

Minority Economic Development

Over centuries federal and other governmental resources have been used by the white majority in America to promote individual and corporate economic advancement. The Black Child Development Institute recommends that governmental child-development resources similarly be used to catalyze economic development in Black and other minority communities and suggests the following as ways to achieve that objective.

JOBS AND CONTRACTING OPPORTUNITIES

Preference for jobs and contracting opportunities must be given to residents in the specific communities to be served by child-development programs if the impact of those programs is to be maximized. The responsible federal official must develop an affirmative action that provides for the participation of contracting firms in the construction, remodeling, renovation, and leasing of facilities and the procurement of goods, services, and equipment. These individuals and firms must be represented in proportion to the total population in the specific geographical areas to be served by government-supported child-development programs.

INVESTMENTS AND DEPOSITS

Legislation should require the responsible federal official to invest and deposit authorized child-development funds in such a way as to strengthen minority financial institutions and otherwise bring the greatest possible economic benefit to families and communities to be served by government-supported child-development programs.

FACILITIES

Provisions for the construction and renovation of facilities and the use of mortgage loan programs must take into account the needs of minority groups, who may not have the initial capital or equity to qualify for bank financing. In order to permit minority groups to compete favorably with other parties, the funding of construction, land purchase, and renovation should be permitted at 100 percent of costs under such circumstances. Ample funds should also be provided for these purposes. In addition federal guidelines requiring a commitment that federally constructed or supported facilities be used for a period of several years should be accompanied by a commitment of operational funds for a similar period of time.

Research

The Black Child Development Institute supports the need for ongoing research that can be translated into programs to foster optimum development of Black children. However, we are concerned that the mechanisms for accomplishing this objective facilitate the acquisition of new and relevant knowledge rather than the solidification of societal beliefs in the inferiority of Black and other minority children.

REVIEW AND EVALUATION COMMISSION

Optimally research conducted by government at whatever level should be monitored by a review and evaluation commission, the majority of whose members shall be representatives of minority groups, parents, the economically disadvantaged, and other target groups served by government-supported child-development programs.

DECENTRALIZATION OF RESEARCH ACTIVITIES

Emphasis and priority should be given to research activities that are sponsored by community controlled child-development centers, the results of which can be incorporated in program content and operations.

POLICY-MAKING AND CONDUCT OF RESEARCH BY MINORITIES

In view of a documented pattern of past exclusion of such input, provisions should be made for the recruitment and selection of minority organizations, institutions, and individuals to occupy policy-making positions and to conduct research projects. This should be done in proportions paralleling the ratio of these minority groups in the target population served by government-supported child-care programs.

COORDINATION AND DISSEMINATION OF RESEARCH FINDINGS

Provisions for the coordination of research strategies among government agencies and the widespread dissemination of research findings among professionals and laymen alike should be an integral part of planning child-development research.

Program Content

COMPREHENSIVE SERVICES

Optimally comprehensive services (medical, educational, social) should be available to all children enrolled in child-development programs to assure the meeting of individual needs and to avoid stigmatizing economically disadvantaged children. Applicant organizations should request funds for these services in their proposals and should make decisions regarding the kinds of services to be provided in that organization in accordance with general federal guidelines.

CULTURE-RELEVANT CURRICULUM

Where children of minority groups are involved, there must be provision of curriculum components that address the unique features of the relevant ethnic and social history, culture, and community life styles, as defined by the specific ethnic or racial groups involved in individual projects. It is also important to acknowledge the right of groups to adapt existing educational models (for example, Montessori) to their own needs or to develop original models relevant to their unique requirements and experiences.

Project Administration

PARENT PARTICIPATION

Under no circumstances should parents be required directly or indirectly to relinquish control over their children to institutions. Pending legislation (Family Assistance Plan) could mandate the participation of children over six in federally supported programs (that is, children of mothers who are the sole supporters of their families). To reaffirm the right of all parents to control their destinies and those of their children, parent participation in the making of policy is non-negotiable. Parents should make up no less than 66⅔ percent of the policy board and should select the balance of the board membership. Parents may also serve in an advisory or volunteer capacity, but this does not obviate the need for their participation in decision-making.

The training of parents so that they may perform their policy-making function should be considered an item for which federal funds can be requested.

STAFFING

1. Racial and ethnic representation on the staff should at the minimum reflect the specific geographic area served, in the same proportions as the incidence of minorities in that area.
2. Priority should be given to the hiring of community residents.
3. State or other certification requirements should not be considered a barrier in recruiting and hiring employees.
4. Recognizing the economic status of most minority communities served by federal programs, the use of volunteers *should not* be mandated, but should be left to the discretion of the local project. If there is volunteer involvement, members of the community served, of all age groups, socioeconomic backgrounds and status, should be considered eligible for service.

TRAINING

The underlying principle regarding training and the use of technical assistance should be to move project personnel to the point of becoming self-sufficient, so that training by outside experts and professionals would no longer be needed.

Individual projects should be responsible for defining their own training needs and identifying training resources.

1. Funds should be available for training both professionals and nonprofessionals in institutional and on-the-job settings with certification provisions that allow for maximum mobility of child-development personnel. (In view of the scarcity of certified and professional personnel, larger percentages of training funds should be provided for nonprofessional training.)
2. Funds should be budgeted for in-service training, and applicants for project grants should be permitted to incorporate funds for that purpose in their proposed budgets.
3. Federal funds for technical assistance should be made available for the purpose of identifying training resources in the community if requested by individual grantees.
4. Once programs are operational, technical assistance should also be provided upon request for the development of career ladders for project employees. Funds should be provided to assist in developing career-development programs geared to local project needs.

OTHER TECHNICAL ASSISTANCE

1. Technical assistance must be made available *in advance* to community applicants for such purposes as developing proposals and interpreting federal guidelines and regulations. Federal assistance should be provided until child-development projects are operational.
2. Applicant organizations must be permitted to include budget items for all kinds of technical assistance in their proposals. Thus projects should determine how such funds should be spent, although they may request regional and federal assistance as needed.
3. Even with more emphasis upon decentralization of the administration of technical assistance funds, a certain percentage of these monies will be dis-

persed from the federal level. There is a need for a national technical assistance plan for child-development programing developed with consumer and community input, which will establish guidelines for the distribution of funds in this area. Contracts and grants let for this purpose should be let competitively as a rule, although there should be sufficient flexibility to permit *sole source* funding of minority firms to meet the specific needs of minority groups. A system for community advertisement of a request for proposal (rfp), particularly among minority firms, is also needed.

ELIGIBILITY

1. Although some economic integration (integration of children of various socioeconomic backgrounds) may be desirable, it *should not* be mandated. Standards should be developed, limiting the participation of economically advantaged children, unless legislation provides funds adequate for care for all children. Consideration should be given to the use of sliding fees in the event that economically advantaged children are enrolled in the program.
2. Economic criteria determining need (for example, standards set by the Bureau of Labor Statistics) should be realistic *for the area served*. Provisions should also be made for periodic revision of income criteria to meet cost of living changes.
3. Geographic location of families to be served *should not* be a factor in determining eligibility. Parents should have the choice of *not* choosing a neighborhood center even though it is convenient.

Federal Standards and Regulations

CONSUMER INPUT

BCDI maintains that Federal Interagency Day-Care Requirements and all other federal standards and administrative regulations for the interpretation of child-development legislation should be developed with the majority participation of parent-consumers.

OVERLAPPING JURISDICTIONS

In the event of conflict between state and municipal standards and federal requirements, federal requirements should apply to all programs funded in whole or in part through federal appropriations. Full participation of parent-consumers and other affected parties should be required when federal standards must be reviewed, delayed, or waived to meet local needs.

Safeguards in the Event of State Involvement

Optimum conditions for the administration by government of child-development programs have already been discussed. Treated in this section are some suggested safeguards that might be utilized should a major state role be legislated.

STATE PLAN

A favorite proposed device for state involvement is the state plan, which provides for the spending of federal monies. The plan is usually to be developed by state authorities, submitted for approval to the responsible federal official, and is then used as the basic guideline for distributing funds.

1. If this approach is used there must be clear federal criteria for approval of state plans. We would recommend that states be mandated, for example, to give priority to economically disadvantaged areas in accordance with the concentration of population and to set aside a certain percentage (for example, a minimum of 20 percent) of the state allocated funds for local community initiated programs (for example, sponsored by grass roots, consumer, and ethnic organizations, corporations, and cooperatives).
2. The acceptance of state plans should also be contingent upon a careful review of the process used to develop the plan. States must have provided for community input and review of the plan prior to its finalization, with sufficient notice and technical assistance to potential applicants.
3. Provisions should be made for amendments to a state plan so that applicants not originally included would not be forced to wait until the next fiscal year.
4. Once the state plan is developed, responsibility for administering funds should ideally remain in federal hands.

STATE COMMISSION

If there is to be a state plan, the recommended vehicle for the development of that plan is a commission that is representative of day-care and community interests in that state. The commission's major function should be to set priorities for funding in the state, following federal guidelines, to amend the plan as the need arises, to serve as a feedback mechanism regarding the effectiveness and adequacy of programs, and to assist in getting communities the help they need. The commission should *not* administer the state's share of federal funds. This should be done by regional federal agencies or municipal prime sponsors. Nor is there a need for a separate state agency to administer funds or to prepare a state plan that the commission then approves.

The composition of a state commission is an important consideration. Optimally a state commission would consist of membership appointed by the governor (including his own representative, public agencies, etc.), membership consisting of established community organizations and private agencies selected by local policy councils or municipal prime sponsors, and at least 66⅔ percent parents selected according to democratic procedures. Minority private agencies and organizations (National Association for Black Child Development, National Urban League, National Welfare Rights Organization, Black Psychiatrists of America) should be encouraged to become part of the commission under these provisions. In addition representation of each minority group on the state commission should reflect at the minimum the percentage of that minority served by government-supported child-development programs in that state.

Provisions must be made *in advance* to inform local communities of the conditions for the selection of the state commission, and technical assistance must be provided to those organizations and groups who seek state representation and involvement. Similarly there should be notification in communities if there are subsequent hearings held by the commission or revisions of the state plan or other issues.

STATE FUNDING

As stated, BCDI maintains that federal funds for child-development programs should *not* be administered by states. Should such a role be mandated, there is a need for an intermediary sanction authority available to the responsible federal official in the event of lack of cooperation, poor performance, and particularly lack of responsiveness to minority groups on the part of the states. Therefore, of those funds to be allocated to states under a designed formula, not all funds should be distributed initially. The balance would be forthcoming when the federal official was satisfied, after a careful review of performance, that the states were complying with federal guidelines and requirements.

Coordinating Mechanisms

CONDITIONS

The establishment of organizations that have as their purpose coordination and cooperation among agencies and organizations involved with child-development programs is advisable if:

1. Adequate funds are made available to affected communities in advance to organize for participation on such mechanisms.
2. The functions of such bodies are defined to permit the individualization of programs according to the needs of specific ethnic and socioeconomic groups.
3. At all levels the representation of parents and community organizations constitutes no less than 66⅔ percent of the total body.

FUNCTIONS

The primary function of such organizations should be the exchange of information concerning child-development activities and program needs, innovations, and the setting of such priorities that are not addressed in existing federal guidelines.

Conclusion

Quality child care for minority group children is not a privilege but a right. Community control is essential, and such programs must be funded at a level adequate for comprehensive care. Research, design, and administration must grow out of the experience of those served by the program, and any child-development center must be approached within the context of *total* community advancement if it is to be effective. These guidelines should be the basis of child-care legislation for all minority communities, regardless of size. Through the application of these principles, the nation would begin to move toward providing quality child care for minority communities.

NOTES

1. The Black Child Development Institute is located at 1028 Connecticut Ave. N.W., Washington, D.C. 20036. Its director and associate director are Evelyn Moore and Maurine McKinley.
2. U.S. Congress, House, Education and Labor Committee, "Expanding Head Start," an Office of Economic Opportunity Report, 1969.

5

YOUNG CHILDREN:
PRIORITIES OR PROBLEMS?
ISSUES AND GOALS FOR THE
NEXT DECADE

PAMELA ROBY

At the moment, child-development services are hotly debated throughout the United States. Persons on the far right attack child-care proposals as measures to "sovietize children" that sound "dangerously like Nazi eugenics," while those on the far left assail them as designs intended to serve the needs of the economy rather than those of women or children. Many others welcome the proposed services as a great stride toward increasing children's and women's well-being and freedom. Following pressure from labor, religious organizations, feminists, minority groups, and educators, the Senate and the House of Representatives passed the Comprehensive Child Development Act in the fall of 1971.[1] Child-care advocates and adversaries waited tensely for the President to sign or veto the bill. Then with a long, passionate veto message Nixon buried the proposal.

Behind this drama, which has been maintained despite the President's action, various goals for child-development services compete with one another, and numerous issues regarding the administration of the proposed programs dance in and out of recurrent policy debates. This chapter will examine the dominant objectives for, and issues concerning, universal, comprehensive child-development services in the United States.

Over recent years child-care and child-development services have been proposed as a means to achieve six primary, partially overlapping, and often conflicting purposes or goals. These objectives, which often are seen in two or three different perspectives, are:

1. Child Development
   a. To fulfill children's right to develop to their full potential during their early years by providing them
      —with an opportunity to enjoy stimulating group experiences,
      —with the opportunity to interact with mothers who are relaxed and fulfilled because they may define how much time they spend with their children and how much time they spend pursuing interests other than their children, and
      —with opportunities for intellectual stimulation to enhance children's conceptual development and to increase their awareness of their community.
   b. To provide the nation with physically and mentally healthy future citizens by providing children with nutrition, health care, and intellectual stimulation.
2. Social Services
   a. To provide the proper developmental environment that will meet the special needs of handicapped children.
   b. To ameliorate the situation of neglected and abused children.
   c. To help troubled families remain intact
      —by sharing with parents the responsibility of care for their children, and
      —by helping parents better understand their children, and
      —by linking parents with other available social services.
3. Women's Rights
   a. To provide a necessary, although not sufficient, condition for women to have full opportunity to participate in America's economic, political, and cultural life.
4. Reduction of Inequality
   a. To provide to all young children health care, nutritious meals, toys, recreation, and education on a sliding scale, with the price of care rising with family income.
   b. To provide a catalyst for the economic development of total communities through community participation and control in the construction, provision of materials, maintenance, and administration of child-development centers.
5. Income Maintenance
   a. To free mothers receiving welfare to be trained to work and to become self-sufficient, and to enable families struggling to remain off welfare to do so through the provision of services.
   b. To provide well-paying training and jobs within child care for members of the community.
6. Female Participation in the Labor Force
   a. To maximize females' contribution to the labor force during the years when they have young children.
   b. To enable women to maintain and further their knowledge and skills during their early childrearing years, so that in their postchildrearing years their contribution to the labor force will not be diminished.

   Wherever and whenever child care is discussed, one finds that not only do numerous issues revolve around each of the goals enumerated above, but that the goals themselves are at issue. We will therefore examine the goals for child-development services in the context of the following highly debated child-care issues: (1) cost, (2) eligibility requirements, (3) revenue base, (4) governance and coordination, (5) staff qualifications, (6) type of program: custodial care or comprehensive child-development services, (7) income maintenance, (8) effects of child care on inequality, and (9) effects of child care on the family.[2] The first six of these issues primarily concern aspects of child-care programs and proposals, while the last three issues relate to the broader social implications of the various proposals.

Cost

How much can America afford to spend on children? Mr. Nixon cited cost as a major objection to the 1971 child-development bill. In his veto message the President stated that "given the limited resources of the Federal budget, and the growing demands upon the Federal taxpayer, the expenditure of $2 billion in a program whose effectiveness has yet to be demonstrated cannot be justified."[3]

Regardless of the setting, one soon finds that the issue of "how much can we afford" serves to disguise the question, how much do we *wish* to spend on young children instead of on highways, defense, higher education, moon shots, and other national priorities, as well as to mask many of the issues discussed below.

As Chapter 8 documents, the cost of full-day, full-year child care varies with what the nation buys. In day-care centers costs are primarily related to the staff-child ratio, staff salaries, and the comprehensiveness of the program. Staff salaries are the largest budget item of child-care centers.[4] The size of this budget item depends upon staff salary levels and the staff-child ratio. Staff salaries, currently often below the federal minimum wage, are rising. The staff-child ratio, most educators agree, crucially affects the quality of care. The amount of attention and love, so important during the early years, that adults can give to children is directly related to the number of children under their supervision. The comprehensiveness of child-development programs refers to whether the programs include health care, meals, transportation, and other supportive services. Each naturally affects the total budget of a child-care program.

Child-development advocates stress that what the nation saves by skimping on child care today may seriously harm many individuals and later cost the society much in remedial health, education, penal, welfare, and manpower training bills. In a survey of current child-care practices the National Council of Jewish Women found infants and young children in abominable settings throughout the United States. Groups of young children were found in filthy basement rooms with rats and broken windows. Others were cared for solely by youth not yet in their teens themselves. In one center licensed to care for no more than six children, one unassisted worker cared for forty-seven children. Eight infants under her control were tied to cribs, toddlers were tied to chairs, and three, four, and five year olds coped as best they could.[5] Hundreds of thousands of other infants and toddlers go to work with their mothers, playing forty hours a week in the back rooms of dry cleaning establishments, sweatshops, and stores. It is difficult to imagine children growing up in such environments as happy, "normal," contributing citizens rather than burdens to society. But more important than the damage that may accrue to society, child-development advocates and humanists stress that human life is our most precious gift and as such should be highly valued. The early childhood years comprise one-tenth of humans' lives, and the dignity, respect, and well-being with which persons can live during those years should be of concern to all.

The two-billion-dollar child-care expenditure vetoed by the President in 1971 represents only one-sixth of the cost of providing care for the children of working mothers. Considerably more would be required to meet the needs of children of low-income families and those who are neglected or handicapped. In coming years the importance of cost as a child-care issue will vary with the political priority given to children and mothers. Today America, the richest nation in the world, stands thirteenth among nations in infant mortality and ranks equally low with respect to maternal, child health, and social services. The creation of adequate funds for child-development services will require the creation of a political priority for children and mothers.

Eligibility

Who are to be the recipients of child-development services? Is child care only for low-income children, working mothers' children, handicapped children, neglected children, four and five year olds but not toddlers and infants? Or is it to be for everyone who wants it? Today, as shown in Chapter 1, child care is available to only a tiny fraction of our children. What should our short-range and long-range goals be for the inclusion of our preschool children in child-development services?

Advocates of child-development services for children from *all* income groups and social classes argue on behalf of both low-income and middle-income children. In Chapter 2 James L. Hymes, Jr., pointed to the benefits that all young children derive from good group experiences with other young children. In Chapter 4 Constantina Safilios-Rothschild reported that young children and their mothers are far more isolated in America than in other societies today or in past cultures. She recommended that women of all income classes be freed from twenty-four hour mothering so that they may become more relaxed, fulfilled mothers and human beings. Such a policy, she suggested, would probably also reduce the numbers of children from all socioeconomic groups who are battered by their mothers.

Others have argued that since our national resources are limited, we should concentrate what funds we have on low-income children and other special groups. In the foreword to this book Congresswoman Shirley Chisholm replies to this argument, stating, "Funds in the United States aren't limited. . . . We scrimp on programs for people because we choose to spend our money on tanks, guns, missiles, and bombs!" After all we can afford planes that cost forty-six million dollars each. Chisholm has also testified concerning other reasons why day care should *not* be limited to the poor:

First, income limitation and means tests are demeaning.

Second, those just over the line, the working poor, those with a toehold in the middle class, and those in the middle class need this resource and service as well as the poor.

Third, we know from our experience with the poverty program that programs exclusively for the poor—no matter how well justified—are not popular. We have seen time and time again how popular resentment has generated enough political pressure so that poverty appropriations are hacked to smithereens on the floor of this House [of Representatives].

All of us are vividly aware of the splits and tensions in this country between the poor and the working class. The lazy bums on the welfare dole versus us middle Americans of the silent majority is the jargon this battle is currently cast in.

Let's not aggravate those tensions. The poor and the working class have the same needs and the same problems. Low wages, inflation, lack of job opportunities, poor educational resources, frustration with the impersonal bureaucracy, and the lack of day-care facilities—they are the same problems. Do not pit these people against each other like starving packs of dogs fighting over the same meager scraps.[6]

Still others have also argued on behalf of low-income children that if middle-income children, whose parents have greater political resources than low-income families, are not included in child-development services, the programs will soon be thwarted by inadequate funding. Speaking for middle-income mothers, on the other hand, Sheila Cole has noted that "the discussion within the government of day care both as a way of 'breaking the cycle of poverty' and as a way to get women off welfare roles has inevitably raised the question with many women who do not claim to be poor: why not us too?"[7] Other middle-income women press for federally subsidized child care for all so that their children may grow up in a socioeconomically integrated setting. "One thing children *do not* need is to grow up in a sterile, homogeneous environment," Vicki Lathom of the National Organization for Women has declared.[8]

Although many argue that universal child-development programs are more likely than selective programs to better the lot of low-income children, others maintain that without explicit attempts to help the poor, children of the well-to-do are more likely to enroll in and benefit from the programs than are children of the disadvantaged.[9] A means test is the most efficient way to direct limited resources to children most in need (see Chapter 6). Some who argue for universal programs as an eventual goal support selective programs for the near future. They maintain that with or without limited funds it is impossible to develop suitable facilities and to train the staff for millions of children within a period of two or three years.[10] However, the creation, within a twenty-four-month period filled with wartime pressures, of nurseries and child-care centers enrolling over a half million American youngsters suggests that the task, although difficult, is not impossible.[11] Today, unlike during World War II, many unemployed construction workers, teachers, women experienced from mothering their own children, and teenagers eager to find some purpose for their lives stand ready to build and run children's centers. Those who need training will have the opportunity to receive it on the job where early childhood growth and development can be observed firsthand.

A final group that believes that public provision of child care should be limited to welfare children consists of those who view child care primarily as a means for mothers to work and for families to get off welfare rolls. Whether the income maintenance goal should be primary is discussed in greater detail later in this chapter.

Discussions concerning eligibility are not limited to the question of family income. The age at which children should be admitted to centers and the hours that centers should be open (which indirectly affects eligibility) are also heatedly debated issues. John Bowlby's findings about the effects of institutional care and maternal deprivation on infants and young children greatly discouraged the development of children's and infant's centers in the United States as well as in England and Wales (see Chapter 23 by Peter Boss) during the 1950s and much of the 1960s.[12] Recently studies have shown that, indeed, two year olds who are placed in residential nurseries where they were separated from their parents from a few weeks to six months—institutions such as those Bowlby studied—do cry more, get sick more often, regress in speech and toilet training, and become more hostile. *But* those who are placed in institutional settings part-day every day and reunited with their parents every evening appeared to behave normally and not to suffer any of the effects that policy-makers generalizing from Bowlby's orphanage studies had expected.[13] Demonstration infant-development programs conducted in the late 1960s and early 1970s (described by Bettye Caldwell in Chapter 3) showed that not only did many parents of all income classes need and want infant-care centers but also that the centers contributed much to infants' cognitive and social development without affecting their emotional adjustment. With more and more mothers not wishing to leave for long the jobs for which they have spent so much time in training, there is a growing demand for quality infant-care centers as well as for centers for toddlers and young children. In addition to this demand parents are pointing to their need for centers open twenty-four hours a day to care for their children for a limited period any time of the day or night. The hours that centers operate determine the population that is able to use them. Approximately half of all persons needing child care require it during times outside the standard work week when most child-care centers now operate (see Chapter 8, p. 115).[14] Millions of families are headed by a single father or mother who must work weekends, nights, or overtime in America's restaurants, stores, factories, or offices (when the daytime staff leaves, the nighttime cleaning, security, and boiler room workers come on). Few or no child-care centers or family day-care programs are available for these families. Their only alternatives are to leave their children alone or in informal arrangements.

Paying for Child Care

Who is to pay for child care? Is it to be "free" for all who wish it, that is, supported through taxes? Is it to be "free" only for the poor and other selected groups, or is it to be provided on a sliding scale with the price of care rising with family income? These questions are related to the issue of eligibility discussed above, but are mentioned separately because they will remain issues after child-development services are made available to all. If middle-income families are to be eligible for services, are they to pay the full cost of the services, receive them on a subsidized basis, or obtain them free of charge?

Advocates of free child care for all argue on behalf of both middle-income and low-income families. Vicki Lathom of the National Organization for Women, has testified about the difficulty middle-income families experience in finding adequate child care that they can afford:

Latest Department of Labor statistics (1970) point to 3.7 million working mothers of preschoolers that are above the defined poverty level. Over 2 million of these are considered traditionally middle-income—that is with family incomes of from $5,000 to $10,000 per year. . . . Children from these mid-income families are too often placed in seriously inadequate and sometimes dangerous child care situations—when they can be afforded or found. A *Good Housekeeping* magazine poll, published in March 1970, showed that the most pervasive complaint of working mothers was the lack of dependable child care. . . . With good developmental care costing around $2,000 per year, it is easy to see that the so-called middle-income parent, as defined above, needs support to receive child care for his or her child.[15]

The National Women's Political Caucus has noted in addition that "the high cost of day-care services" as well as "the lack of day-care facilities has forced many women onto the welfare rolls."

Persons speaking on behalf of low-income groups argue that means tests, which would be required if the cost of child care were to be contingent on family income, affront human dignity and that their stigmatizing effects may discourage many persons who qualify for service from using them. On the other hand the Black Child Development Institute (Chapter 6) and others have argued that means-tested child care is better than inadequately funded child care. They urge that sliding fees for economically advantaged children be considered unless the legislature provides funds for quality child-development programs for all.

Governing

The creation of nationwide child-development services necessitates the creation of a coordinating apparatus (if not a bureaucratic monster) and raises issues of power. Who is to control the programs? Who is to decide the nature of the influence they are to have upon the children in them? Parents? The community? Professionals? State or city politicians?

Governing issues span two major concerns: (1) what should the role of parents and local communities be in policy-making; and (2) how should the federal government coordinate and disburse funds to federally funded child-care programs. In this section these shall be discussed separately.

COMMUNITY AND PARENTAL CONTROL

Minority group representatives, city politicians, feminist organizations, and political coalitions formed for community control of schools all want child-care programs to be locally controlled. But what *is* local control? Is it governing by persons living within an area served by an individual child-care center or perhaps two or three such centers? Is it direction by communities of 5,000 or more as proposed in the Mondale bill, by SMSAs (Standard Metropolitan Statistical Areas) of 50,000 or more, or by cities of over 100,000, as proposed in the Brademas bill (see Chapter 5)?

Minority representatives and radical groups demand that they be allowed to set policy for centers in areas where they are a majority. Every large city of the nation has Black, Puerto Rican, or Chicano ghetto communities. Residents of these communities wish to make their own policies concerning child-development services for their children. If the boundaries for "local" control of child care are drawn according to city, SMSA, or county boundaries, rather than along the lines of smaller "communities of interest" including those of ghettos, ghetto residents will again be without a voice in decisions significantly affecting their lives. The nation does not have one set of social values or one life style. It has many. Local control at the community level where persons share somewhat similar values is the only way to avoid one group being oppressed by another. Blacks and other minorities have learned that institutions nominally serving them but controlled by the white majority may serve the needs of whites, but do not serve their needs (see Chapter 6 by the Black Child Development Institute and Chapter 16 by John Dill). Maurine McKinley of the Black Child Development Institute has testified, "The desire for community control of day care and child development programs is born of the realization that the first five years of life are too critical to be entrusted to the exclusive molding of those for whom the black child's interests may not be predominant." [16]

Radicals and feminists have also stated the issue poignantly: With "the hidden curriculum" of most centers, "by the age of four, children are assimilating the idea that a woman's place is in the home" and "that it's better to be white. They are learning to follow directions without asking why. They are learning how to deny their own feelings and needs in order to win approval from adults. . . . As radicals we must understand that our goals for children are in conflict with those of the institutions—corporations and universities." Therefore, "control must rest with those who struggle for and use the day care centers." [17]

Conservatives too seemingly should be in favor of community control. However, they have opposed national child care altogether rather than favoring community control. The conservative weekly *Human Events* described the 1971 Senate- and House-passed child-care bill as a measure "to remove the education and training from the home and the church and turn it over to an agency of the federal government," and listed the eighty-one Republicans who voted in favor of the bill under the banner "RENEGADE REPUBLICANS." [18] Russell Kirk described the bill as "a colossal design for making as many little children as possible the wards of the federal government," and wrote:

> Never before, in my decade as a columnist, have I urged readers to write to the President, asking him to veto a piece of legislation. But I so urge you now. Write to Nixon today, if you believe that a parent is the proper guardian of a little child, not the national government. [19]

As these writers suggest, issues concerning parental control are akin to those of local control. At the 1970 White House Conference on Children and Youth, the Task Force on Delivery of Developmental Child-Care Services recommended that parents should control programs whenever feasible and that at least one-half of the places on the governing boards of publicly funded programs should go to parents. [20] Others, believing that parental control is the best mechanism through which to ensure quality programs, recommend that parents or parents and community residents comprise the entire board. [21] What are these boards to do? What are community and parental involvement to mean? As Don Miller's categorization of four levels of parental involvement in Chapter 7 shows, parents' relationship to children's centers may vary from giving a bit of advice to teachers with no guarantee that it will be heeded to having parent and community controlled boards that select center staff and employ professional consultants to work for parental interests. The Delivery Task Force of the White House Conference recommended that "parents of enrolled children must control the program at least by having the power to hire and fire the director and by being consulted on other positions," and that "parents and local communities must also control local distribution of funds and community planning and coordination . . . and play a role in the flexible administration of standards, licensing and monitoring" of programs.[22] Boards were not to be advisory but policy-making bodies.[23]

The role professionals should play in a parent and community controlled system is also at issue. Although each professional group has its own perspective, many from each discipline shudder at the thought of parental control and actively oppose it. They claim, "professionals have been educated in what is best for children; parents, while emotionally concerned, have not." Increasingly parents and communities are responding that all too often the rhetoric of professional expertise is used to cloak the personal values of professionals. These parents want the expertise of professionals who will work *for them* and who will use their expertise in service of their community's needs, values, and interests (see Chapter 16).

Opposing advocates of parental and community control are not only spokesmen for professional interests but also persons concerned with racial and class integration, administration, redistribution, political feasibility, and maximum utilization of the states' inputs into child care. Community control, many liberals fear, spells the end of racial and class integration. Integration, they maintain, is a prerequisite for the children of the poor to learn the language, manners, and life style required to enter the mainstream of the American economy. Many minority and ghetto community representatives counter this argument by saying that the economy and social system must adapt to their needs, values, and culture; that it must allow them to be successful on their own terms.[24] Ruth Turner Perot, organizer and former director of the Black Child Development Institute, states, for example, "What we have to recognize in America is that, one, black vitality and creativity cannot be nurtured in white institutions. Two, we must make it possible for institutions to develop in which blacks can search for their own routes to liberation. And, three, it is only on this basis that effective partnerships can be made with others." [25]

Others argue for federal control because they believe that resources are likely to be distributed very much in favor of the rich *unless* control over the distribution of resources and staff is placed at the federal level away from the political pressures of local businesses and wealthy community residents. This argument has logical merits, but looking back a few years, we see that even the federal Elementary and Secondary Education Act (ESEA) Title I funds, which were intended solely for the in-classroom use of low-income children, were in fact seldomly used for that purpose.[26] The lessons of the ESEA would indicate that the federal government cannot effectively monitor funds from Washington. Professional representatives of communities as large as school districts spend money earmarked for the poor on children of the well-to-do or on local enterprises that can reward them politically. The lessons also suggest that parental and community control over the distribution and use of funds within communities, coupled with strict federal control over the distribution of funds to communities, is the best insurance that funds will be spent to the advantage of low-income children and community residents.

The Nixon administration objected to a similar proposal embodied in the 1971 child-development bill, saying that it would create an "administrative nightmare" [27] Others maintain that community control, whether administratively practical or not, is simply not politically feasible. They point to the experience of Community Action programs and the concept of maximum feasible participation in the inner cities; militant Community Action agencies either were eventually taken over by established institutions or ended up in bitter conflicts with them, resulting in city mayors and other politicians pressuring Washington to hand control back to them.[28] Still others, clearly speaking for those who benefit from the status quo, contend that the nation needs to be united and that community control of child care and other institutions will only serve to further divide it. Those who speak for community control would also like to see the nation united, but their terms and the terms of those who speak most vociferously for a unified nation quite obviously differ.

Finally, related to the issue of local control is the whole question of states' roles in the delivery of child-care services. States are already considerably involved in early childhood programs, and as Sally Allen points out in Chapter 12, their interest is growing. State governments now contribute sizable funds to kindergartens and child care. Supporters of states' playing a significant role in early childhood programs note also that they are a source of expertise and have pooled their information and professional resources on early childhood development by creating the Task Force on Early Childhood Education under the Education Commission of the States.

On the other hand the Black Child Development Institute and groups representing other minorities strongly oppose state involvement in the administration of federally funded child-development programs (see Chapters 6 and 16). While minority representation at the state level is scanty in most states and directly opposed in other states such as Mississippi, *local* control enables minority groups to set policy in areas where they live.

Both those who feel positively and those who feel negatively about the states' role in early childhood programs agree that state politicians have significant strength and that established state early childhood bureaucracies, like any bureaucracy, will be difficult to dissolve, ignore, or prevent from growing.

FEDERAL COORDINATION

The federal government has over sixty different funding programs for child care or child development (see Chapter 11.[29] Each piece of federal legislation creating this funding maze built a separate vertical delivery system. Each system has different goals, little horizontal exchange of information between it and other programs, a different category of eligible clientele, separate agencies, different procedures and guidelines, and different geographic boundaries defining local communities for planning and service delivery.[30] Initially each piece of federal legislation was created to meet the needs of a separate interest group and government bureaucracy. Owners of large corporate farms and their representative, the Department of Agriculture, had a food surplus that was reduced through the National School Lunch Act and the Child Nutrition Act. The need for low-skilled female workers and the desire to reduce welfare rolls led to the inception of the Work Incentive Program (WIN) and the Concentrated Employment Program (CEP) under the Office of Social Security and the Department of Labor. Children's needs were secondary for these program planners.

The establishment of child-care programs for multiple, conflicting purposes has created a chaotic, uncoordinated nonsystem that serves children poorly if at all.[31] The nonsystem is characterized by competition and conflict among agencies. Secrecy rather than cooperation and sharing of information is the norm at all levels. Gwen Morgan reports one result:

> At this writing, two family day care programs are being planned in a single housing project by two groups which are planning separately, one of them relating to a Mental Health orientation and the other to a Welfare orientation. Residents are being manipulated by professionals, and allow this to happen in the expectation that both projects will be funded. Meanwhile other parts of the city have no services.[32]

Another characteristic of the nonsystem—federal agencies' defining local communities by different geographical boundaries—compels each agency to collect its own data and makes exchange of data among local groups purposeless. This characteristic also prevents community people from having one board or council to oversee all child-care programs in their neighborhood. The nonsystem is confusing to local citizens and professionals alike and results in the duplication, triplication, and quadruplication of efforts at all levels of the child-care hierarchy. Such duplication is financially costly as well as wasteful of concerned persons' time. Each program within the nonsystem has its own proposal forms, procedures, responsible persons, and deadlines. In order to obtain funding for the coming year, the directors of child-care centers must either spend most of their own time in figuring out and filling out proposal forms and entertaining representatives of the respective federal agencies—time that they cannot then spend planning with or helping staff—or devote a large portion of their center's budget to hiring a proposal writer or grant person. In either case children are the losers. The inordinate amount of time required by this duplication also prevents parents from actively participating in funding procedures and restricts program formulation to professionals and top federal officials.

The chaotic nonsystem has at least two other negative effects upon children. First, it separates children categorically: welfare children must attend one program, handicapped children another, the children of nurses yet another, and so forth. Rather than being separated, most specialists assert that each of these groups of children needs to live and learn from one another. Second, the mutiple federal and state child-care agencies result in highly specialized child-care programs rather than programs designed to serve the total needs of the child and his family —either grants for one type of program restrict centers from seeking other program funding, or centers simply run out of the steam needed to write more proposals.[33]

Obviously needed is a vastly reduced number of federal and state agencies dealing with child care. After being forced to work with literally dozens of child-related agencies, today's child-care advocate might be tempted to suggest that the agencies be reduced to one. But many caution that this might not be a happy solution.[34] All checks and balances would be gone. Parents, professionals, and community members would have to place their trust in one federal agency. If that agency's actions disturbed them, they would have no other place to go. For this reason perhaps there should be not one but a few—and certainly not sixty-one—agencies for children.

Agency consolidation and coordination would reduce per child child-care costs through broader sharing of services, joint purchasing, and elimination of overlap. Given the number of children not yet covered by child-development services, the qualitative improvement that child-care services require, and the number of services that must be added to most child-care programs before they meet the total needs of children, the desirability of such saving is undebatable. Agency coordination is also needed to synchronize the geographical boundaries by which local communities are defined, so that community boards may develop, oversee, and coordinate the various agencies' child-care programs and pool information that they gather. Furthermore, agency consolidation is needed so that parents and community members may be relieved of having to become full-time experts in red tape before they can shape policies affecting their children. Finally coordination of federal and state child-care agencies is a prerequisite for equitable distribution of child-care resources, for no matter how "equity" is defined, it cannot be obtained until the various funding agencies know to which groups and communities each is disbursing its funds.

Each of these factors points to the necessity of coordinating and consolidating child-care programs. Politically the task will not be easy, for although each agency and interest group recognizes the need for coordinated child-care programs and policies, each wants to be the one to coordinate the others.

Staff

The issues revolving around the staffing of centers can be summarized, if not solved, fairly easily. What characteristics should child-care workers have? Are they to meet specific professional standards? If so, what? Are they to reflect the ethnic and racial characteristics of the population? If so, how is this representation to be obtained?

On one side professionals and their respective associations can be heard urging that all federally supported child-development centers have "adequately trained professional staffs" and that institutions of higher education develop preparatory programs and "encourage individuals to pursue degrees in early childhood education." [35] These professionals generally believe that relatively untrained paid paraprofessionals or volunteers and a career hierarchy are useful concepts. But they emphasize the need for paraprofessionals to work toward professionalization and for all child development center staffs to include persons certified with at least a B.A. and preferably an M.A. in early childhood education. Aligned with early childhood professionals are program directors, who realize that although a university credential may not be the best determinant of ability, it is the easiest means of selecting personnel and is perhaps the least likely to be challenged.

Minority and feminist representatives stand on the other side of the question. They maintain that whatever the merits of credentials, the attainment of a racially and sexually balanced staff is a more important consideration. In Chapter 6, the Black Child Development Institute urges that priority should be given to the hiring of community residents and that at the minimum the racial and ethnic composition of the staff should reflect the specific geographic area served. Elsewhere Evelyn Moore, director of the Institute, has stressed the need for research on programs for Black children to be conducted by Black researchers. She has testified:

We can no longer let white researchers continue to describe, define, and program for black children, because we are finding that their interpretations are leading us down a dead-end street. For example, there are some people who want to do away with aggressions in children who are hyperactive and many of these children are black. Now, aggression in a white child might be interpreted as self assertion of leadership potential. But when viewed in the context of a black child, he is hostile, he is negative. Then we go to the other end of the continuum and we take the withdrawn child. If he is white he might be viewed by the whites as a deep thinker, but you take a black child and he is viewed as apathetic. He might be viewed as even being retarded. . . . I don't think this eliminates white researchers on the basis of any technical assistance that they may be able to offer. But I do not think they are able to view the child, a black child, from the black perspective, by virtue of their own experiences. That is not an attempt to put down white researchers, but just to put them in the correct slot.[36]

The National Organization for Women and other feminist groups seek policy provisions ensuring equitable employment of women and men in the child-care field. This requires more men relating to children in the preschool setting as well as more women participating in administration of the overall program. They argue that men must be a part of preschool staffs, so that children avoid developing at a young age the sex-role stereotype that child care is a woman's job.[37]

Still others maintain that professional credentialing of child-care workers has a negative rather than neutral or positive effect upon what happens within a child-development center. Somehow, four or five years within an institution of higher education makes a person better able to research children than to relate to them with patience and warmth. For this reason some child-care center directors report that they are more inclined to hire a woman with the experience of rearing her own children and only three or four years of high school than a person with a B.A. or M.A. who lacks this experience.[38] Others charge that professional degrees unnecessarily increase the social distance between the staff and parents without degrees.[39] Not credentials but parental control or involvement in staff hiring is the best guarantee of quality, these persons contend.

Custodial Care or Comprehensive Child-Development Services?

The child-development programs provided for in Title V of the 1971 Office of Economic Opportunity bill are "not *just* a babysitting operation to provide custodial care for children while their mothers work. The bill emphasizes the well-being of children and the comprehensive services they need for full development, whether their mothers work or not," Alice Rivlin observed in the *Washington Post*.[40] The fact that Title V did indeed propose comprehensive child-development services subjected the bill to virulent attack by conservatives and ultimately to veto by the President.

What is comprehensive child development? Child-development programs are most often contrasted with custodial child care. Developmental programs address the total needs of the child—physical, social, emotional, and intellectual—and his family in order to enable him to realize his fullest potential. They have a small child-staff ratio, parent involvement, well-trained and well-paid personnel, and good facilities. Developmental programs must be comprehensive programs that not only watch over and educate the children involved but provide them and their parents with medical, nutritional, and social services. Custodial programs, on the other hand, purport to give children good care but have no special developmental or educational component. In practice children do learn when they are in *good* custodial care, and many programs that purport to be developmental are nothing more than good baby-sitting operations with a few minutes of educational games thrown in.

In policy discussions concerning child care, the term "child-development programs" has come to stand for higher quality child care and "custodial care" for lower quality care. In addition to parental and community control discussed above, quality has been measured primarily in terms of the staff-child ratio, with the rationale, as Mary Rowe and Ralph Husby note in Chapter 8, that if at the very minimum we seek child care that is equivalent to good home care, we must take into consideration that most homes have fewer than four children per adult. Child-care center funds are spent primarily on staff salaries. Low per-child costs mean either that the center must have a large number of children per child-related adult or that it must pay its staff very low salaries. Currently many persons working full-time in federally sponsored child-care programs "earn" a poverty level income. The average salary of full-time child-care teachers (including head teachers and teacher supervisors) in a national sample of child-care centers surveyed under OEO in 1970 was less than ninety dollars per week. The average wage of full-time teachers' aides was under seventy-five dollars.[41] These low salaries represent our exploitation of child-care workers. They also indicate our level of concern for children. Child-care workers who are paid poverty wages will often necessarily be more concerned about their own financial problems than about the children with whom they are working.[42] Not only is it impossible to expect child-care workers living under financial oppression to provide quality developmental care, but also it is unrealistic to expect them to provide good baby-sitting services. The key to the provision of quality developmental care will be the level of funding that the nation is willing to provide for children.

Conservatives attack developmental child-care services as a means for "indoctrinating little children and propagandizing their parents." Many maintain that this attack creates a false issue. They argue that parent- and community-controlled child care would be a step toward parents' retrieving control over children's thoughts from educational toy industries and centralized children's television series. That parents do not fear the "thought control potential" of child-care centers is evidenced by the nation's private nurseries for two to three year olds, its upper-middle-class suburban kindergartens for four and five year olds, and its day-care centers for children of the poor. The question that faces us today is not whether the nation will have child care, but whether it will have horrendous, adequate, or excellent care.

Persons who seek high-quality, comprehensive child development do so for the well-being and happiness of children and for the future well-being of the nation. In Chapter 2, James L. Hymes, Jr., stressed that children need a good group experience. In addition to this experience many children need, but live in families that can ill afford, health and other social services. The 1970 White House Conference on Children and Youth emphasized the need for such programs by ranking "comprehensive family-oriented child development programs including health services, day care and early childhood education" as its number one recommendation and overriding concern for the nation.[43] Therese Lansburgh summarized well the feelings of most participants in that conference:

> Do we want our children to have all the benefits of what the behavioral scientists have discovered to be best for them or do we remain in a season of our past, our covered wagon pioneer days? . . .
> Do we want to still consider child care as children in basement rooms staring at television for most of a day that, under expert guidance, could have been a day of the joy of growth and learning? [44]

Although in no way wishing to de-emphasize the importance of child-development programs, John Dill and others have stressed that the programs cannot be expected to accomplish everything (see Chapter 16). To expect them to do so (as Head Start was expected to do) would not only be misleading but harmful to low-income groups. The provision of eight or nine hours of good developmental child care plus good health care and nutrition will *not* give children of the poor a "head start" or even an equal start in life. For equal opportunity to be a reality in this nation, children must begin with equal conditions. Comprehensive child-development programs grant a bit of sunshine to children whose parents are burdened by bad health caused by lack of medical attention; whose own health is not what it should be because as infants they did not have checkups and shots that are taken for granted among middle-class babies; whose psychological and emotional stability is shattered by fears of rat attacks or by their parents' constant worries about basic finances and the junkie down the street. Developmental programs cannot grant children living in these circumstances equal opportunity no matter how bright or motivated the children may be. To condemn these children for not flourishing in school or on the job after attending a child-development program that is only a fraction of their total experience only reveals our insensitivity to their real-life circumstances.

Income Maintenance

WORK-FARE

H.R. 1, Opportunities for Families Program, Section 2101, contains . this statement:

Purpose:
—providing for members of needy families with children the manpower services, training, employment, child care, family planning, and related services
which are necessary to train them, prepare them for employment, and otherwise assist them in securing and retaining regular employment and having the
opportunity for advancement in employment, to the end that such families will
be restored to self-supporting, independent, and useful roles in their
communities. . . .

The Opportunities for Families Program proposed by the Nixon Administration embodies the goal of many for child care: enabling welfare
mothers to become self-sufficient in order to reduce taxes. Work is primary; child care is secondary.

The work-fare proposal would provide child care for families who
need it to work or to attend a manpower training program in order to
get off welfare, but would provide it neither for women who need it in
order to continue working and to stay off welfare nor for women with
incomes as small as $4,300 for a family of four who work to maintain
their families and can only afford inadequate child care for their children.

Under H.R. 1 (Sections 2102 and 2111) families with incomes below
$2,400 can qualify for Family Assistance Plan payments only if all em-
ployable members have registered with the U.S. Employment Service for
manpower services, training, and employment. Mothers whose children
have reached age three (or until July 1, 1974, age six) are included as
employable. The Department of Labor is to furnish child care (Section
2112) for mothers who need the services in order to accept or continue
to participate in manpower services, training, or employment under the
Opportunities for Families Program. The Department of Health, Education, and Welfare is to do the same (Section 2133) for mothers with a
temporary incapacity for work who need child care to participate in vocational rehabilitation.[45] The Secretary of Health, Education, and Welfare is to prescribe the extent to which families are to be required to pay
the costs of the child care; to establish, with the concurrence of the Secretary of Labor, standards for the quality of child-care services provided
under H.R. 1; and to coordinate the provision of this child care with
other child-care and social-service programs (Section 2134a).

The most important question for those favoring child care as a means
of reducing welfare taxes is can it do that. Economist Gilbert Steiner
doubts whether day care will reduce welfare costs. First of all, he points
out, 43 percent of mothers currently receiving welfare have less than a

ninth grade education. For this undereducated group work training leading to employment at wages adequate to support a family is likely, at best, to be prolonged and to result in high child-care and training bills for the taxpayer.[46] Secondly, he notes (as Mary Rowe and Ralph Husby did in Chapter 8), child care is expensive. Third, he concludes, after a few years "it will inevitably be discovered that work training and day care have had little effect on the number of welfare dependents and no depressing effect on public relief costs. (After all, where are jobs for all these women to come from?) Some new solution will then be proposed, but the more realistic approach would be to accept the need for more welfare and to reject continued fantasizing about day care and 'work-fare' as miracle cures." [47]

The second question concerning this form of child-care program involves values: is it ethical to *force* mothers with young children to work outside the home? David Gil and groups advocating mothers' wages argue that society should recognize and compensate mothers for their assumption of the complex and physically tiring task of rearing the nation's future citizens.[48] The Welfare Rights Organization (WRO) and feminist groups rail against the proposed use of force—for example, withdrawal of all welfare payments to a family—to make a woman accept what in all likelihood will be an exhaustive, boring, low-paying, exploitative job for herself and potentially inadequate child care for her child. They assert that in a nation as affluent as ours, welfare is a right. If a "welfare mother" had inherited land from a rich uncle or sugar daddy, she would, along with Senator Eastland, John Wayne, and the DuPonts receive thousands of dollars for letting the land lay idle.[49] If she had "served her nation" by joining the army and killing and terrorizing other human beings rather than bearing and rearing its children, her right to veterans' benefits would be unquestioned.[50]

It is probably impossible to force mothers to work for long. As every employer knows, employees have innumerable means by which they can feign illness or make their presence undesirable and thereby be fired or "let go." Once an individual has established a record of bad work behavior, it is unlikely that the Employment Service will be able to find work for her.

A study of welfare mothers analyzed by Jessie Bernard suggests a possible policy solution for both issues: whether child-care costs will outweigh former welfare costs and whether women can be forced to work. Given good child care and the availability of a good job, mothers' willingness to work, she found, is directly related to the number of young children they have. A majority of women with only one or two children would like to work; a smaller fraction of those with large families (for whom child care would cost the taxpayer a great deal) would be willing to work.[51] Thus, if we were to provide good child-development services, and allow mothers the freedom to choose to join or not join the labor force, their decisions would be likely to be more favorable to taxpayers

than would either the policy of providing no child care and work opportunities for welfare mothers (forcing them to remain home) or the policy
of forcing them to work.

In addition to the questions examined above, administrative policy
discussions concerning H.R. 1, as well as the wording of H.R. 1, provoke
concern about standards for child-care facilities.[52] H.R. 1 (as drafted in
May 1971) says little about standards for child care. It states only: "In
order to promote the effective provision of child-care services, the Secretary of Health, Education, and Welfare shall establish, with the concurrence of the Secretary of Labor, standards assuring the quality of child-
care services provided under this title. . . ." (Section 2134a). Nowhere
does the bill provide for parent or community participation, even in the
most basic forms, in the governing of and decision-making concerning
child-care programs. Nowhere does it describe the procedure a mother
is to follow if she objects to the child care furnished under the program.
And nowhere does it state that a family may choose for itself the best
child-care arrangements and that their choice need not be based solely
upon cost considerations. Not only does the proposal omit mention of
parental and community control, but also it does not define "quality
child-care services."

Persons concerned with children's development would feel more com-
fortable with the H.R. 1 section on standards quoted above if it stated
that "the Secretary of Health, Education, and Welfare shall ensure that
each child-care program under this act meets the Federal Interagency
Guidelines and that each child under this program receives the educational, health, nutritional, and related services he needs to help him
achieve his full potential." These omissions, coupled with the limitation
of H.R. 1 child care to welfare families and the mandatory work requirement, reflect the philosophical goal of the bill, which is to put welfare
mothers to work and reduce taxes. Child development is not even a secondary goal of the bill; the very wording of H.R. 1 indicates that
child development is not an aim of the bill at all. Therese Lansburgh,
when president of the Day Care and Child Development Council of
America, phrased the problem well:

"A 'good' mandatory day care program is a contradiction in terms.
In a free society, the parent has both the responsibility and the right to
decide what is in the best interest of her children." [53]

In addition HEW internal memoranda propose very low dollar
amounts per child for full year child care: $1,600 for preschool center
care, $730 for center care (summers and after school) of school-age chil-
dren, $894 for preschoolers and $716 for school-age children cared for in
their homes, and $866 for preschool and $542 for school-age family day
care.[54] These amounts are far beneath what has been found to be the
cost of acceptable child care, let alone desirable child-development programs (see Chapter 8).

The final aspect of H.R. 1 that suggests that it is a child-retardation rather than a child-development bill is the very low benefits it provides to families without an employable member. A family consisting of three children and a mother certified as unemployable would be provided with a total yearly income of $2,400. This amount is $1,568 below the 1970 poverty level of $3,968, defined by the Social Security Administration as the minimum amount necessary for subsistence, and far below the $6,500 that the Bureau of Labor Statistics has set as the approximate level of income necessary for a family of four to have a living standard commensurate with an adequate diet. Under H.R. 1 no family may receive food stamps with its cash benefit. As of July 1970 Columbia University's Center on Social Welfare Policy and Law reported that only five states paid less than the H.R. 1 benefit level in AFDC and food stamp benefits. Over 89 percent of all AFDC recipients, or 7,836,700 people, live in the forty-five states that currently provide benefits higher than those proposed in H.R. 1.[55] Children raised in the abject poverty provided under this program can hardly be expected to grow to be healthy, contributing adults. Rather, by saving the taxpayer a bit of money today, the proposal, if enacted, will create circumstances leading to an ever growing population of adults who are either mentally retarded and physically weakened by malnutrition or undereducated and emotionally unstable as a result of it. If we as a nation cannot respond to this deprivation in human terms, we should realize that unless we, like Nazi Germany, turn to exterminating our own citizens, the economic costs of maintaining such an adult population a few years from now will far outweigh our immediate savings.

COMMUNITY DEVELOPMENT

Good child-development programs *can* provide income maintenance and contribute to community development (see Chapter 6 by the Black Child Development Institute). The creation of child-development services does, after all, require hundreds of thousands of additional professional and paraprofessional child-care workers, cooks, administrators, maintenance workers, and construction workers as well as equipment for the centers. All of these can come from and bring money into low-income communities.

A good child-development program that is *also* a good income maintenance and community-development program would, for the sake of the children involved, for the sake of the workers, and for the sake of community development, pay all its staff adequate salaries—something well above the minimum wage. It would, unlike most urban school systems and federal programs today, "buy Black," buy Puerto Rican, buy Chicano, and buy inner city. Not only its lower echelon personnel, but also its middle and top level administrators in local, regional, and federal offices would—unlike those in the Office of Economic Opportunity, HEW, and Labor Department programs today—come primarily from low-income communities.

The program *would* provide many of today's poverty level families with adequate incomes. It *would* serve as a catalyst for the economic and social revitalization of low-income communities. And it might begin to reduce the divisions and crime within this nation.[56]

Child Development Programs and Inequality

Many Americans' attempt to grapple with poverty during the 1960s led to renewed recognition of the prevalence of inequality in affluent societies.[57] Absolute increases in living standards are important. They affect individuals' physical health as well as satisfaction. But individuals judge their level of living not only in comparison with what they had ten, twenty, or thirty years ago, but also in comparison with the standard of living around them. The sting of poverty in an affluent society is one's comparative lower standard.

Children, unlike their parents, have *only* the current comparative perspective. It is difficult and unrealistic to expect the children of the poor to have a "good-self image" and to believe that the nation really cares for them—that they should try to contribute to rather than "rip off" their society—as long as every day and every hour they see that more well-to-do Americans fail to put their money where their mouths are. Although the nation's leaders state that we must come again to respect every person's work no matter what that work may be, in America money and material things, which our society values so highly, are distributed so that many full-time workers do not receive enough to adequately feed and clothe their families while others who work less live in affluence.

In 1970, 5.5 percent of the nation's family income went to the bottom quintile of the family population (that is, to the poorest 20 percent when families are ranked from richest to poorest), and 41.6 percent to the richest quintile.[58] Between 1947 and 1970 (the period for which we have annual census statistics) little progress was made in increasing the share of the poorest 20 percent of our nation's families. Over the twenty-three years total money income going to the lowest quintile of families rose only one-half of 1 percent: from 5.0 to 5.5 percent.[59]

What impact are child-development programs to have upon the distribution of income and well-being in America? The compensatory education, manpower training, and Head Start programs of the war on poverty years were all based on the premise that education is the best strategy for reducing poverty and inequality. Today when poverty is found to be still with us, it is often argued that the programs didn't start with young enough children or that the children's families or genes were

somehow at fault. If these arguments stood up, child-development programs would be an important means of reducing inequality. Unfortunately, as important as child-development services are for improving
children's day-to-day well-being, the services are unlikely to affect children's future economic status or opportunities unless the American economic and social structure is significantly changed by other means.

First, as noted above, for low-income children to have an equal start
toward obtaining a good education and a well-paying job, they need not
only quality child-development programs but also good housing, safe
and accessible outdoor areas in which to play, adequate and stable family incomes, good nutrition and health care for their parents and themselves, and societal respect for their culture and family. Second, even if
we should radically extend our political imaginations and produce all of
the above, youth would still be faced with an economic system in which
the income structure is highly unequal. In this economic system the income people receive has very little to do with their education and effort.
Instead the variance between persons' wages is accounted for primarily
by markets for labor—their rates of growth and rates of unemployment;
discrimination on the basis of race, ethnicity, age, sex, and social class;
and government activity.[60] One in four of all families classified as poor
by the Social Security Administration in 1966 was headed by a man who
worked throughout the year. The families of these men who "earned"
their poverty included eight *million* persons.[61]

Furthermore, in this economic system the incomes of the rich depend
upon income from work—wages and salaries—much less than do those
of the poor. Taxpayers with net taxable incomes exceeding $100,000 re-
ceived only 15 percent of their income from work and 67 percent from
dividends and capital gains (the remaining 13 percent came from "small
business" income); those with net taxable incomes under $20,000 re-
ceived 87 percent of their income from wages and salaries and only 3
percent from dividends and capital gains in 1966.[62]

Although child-development services are unlikely to change the income distribution, they could be used directly to shift other resources
from rich to poor. A federally sponsored child-development program
could provide *all* children with adequate nutrition, good health care,
and stimulating educational experiences, as well as provide jobs for the
many unemployed parents who wish to work, and it *could* be financed
through a program of progressive taxation.[63] The Comprehensive Child-
Development Program vetoed by Nixon might have redistributed resources and thereby reduced inequality even if ever so slightly. The
Welfare Reform Plan (H.R. 1), by requiring mothers to register for work
in order to obtain monthly income benefits of approximately $200 (the
proposed amount for a family of four), is likely to flood the low-wage
market (where unemployment is already high), and thereby to further
depress wages for both men and women and increase inequality.[64] In
the short run this policy would benefit employers and hurt workers. In

the long run it is likely to hurt the entire nation by creating greater bitterness and division.

Although increased differentials between the wages of high and low paid groups can be almost guaranteed to result from the implementation of the Nixon administration's Welfare Reform Plan as currently constructed, increased inequality is *not* a necessary consequence of child care. Although the provision of child care and other social services by themselves is unlikely to affect economic inequalities, income inequality *can* be reduced or increased by direct manipulation of the economy through taxation; wage supplements; income transfers (similar to veterans' benefits, welfare payments, and farm subsidies); government creation of well-paying jobs in child care, education, housing, construction, park renovation, and other areas where personnel are badly needed to improve general levels of well-being; and wage or profit freezes (the 1971 wage freeze severely affected the incomes of persons earning under $20,000 while barely touching those of persons in the over-$100,000 income bracket). To reduce inequality through direct manipulation of income will require a political will. But the experience of the past decades suggests that only with this will and with direct restructuring of the income distribution will the goal of inequality reduction be realized. Not until then will all children have a future to which they can look forward.

Child Development and American Family and Religious Life

What impact does our family-centered childrearing approach have upon children and parents? What impact would we like our relations with children to have?

The Nixon administration's position on the family was made clear in the President's message vetoing the Comprehensive Child Development amendment to the 1971 OEO bill:

. . . . all other factors being equal, good public policy requires that we enhance rather than diminish both parental authority and parental involvement with children—particularly in those decisive years when social attitudes and a conscience are formed and religious and moral principles are first inculcated. . . .

. . . for the Federal Government to plunge headlong financially into supporting child development would commit the vast moral authority of the National Government to the side of communal¹ approaches to child rearing over against the family-centered approach.

This President, this Government, is unwilling to take that step.[65]

Groups responded quickly to the administration's statement. Before the House of Representatives Congressman Drinan proclaimed that "the President's veto of the child development program undermines the

American family" and reminded the nation that the proposal had been endorsed by both the National Council of Churches and the United States Catholic Conference.[66] A *New York Times* editorial wrote,

> The President's charge that day care weakens the family ignores the realities of much of modern family life. Poor and working-class families normally have to leave their children improperly supervised or entirely unattended for much of the day; families at virtually all other income levels rely heavily on baby-sitters and, in the upper brackets, on a variety of domestic help.[67]

Others noted that contrary to the implications of the President's veto message, communal child-care approaches are not new to the United States.[68] As Virginia Kerr describes in Chapter 10, they have a long history.

Questions concerning childrearing and American family life extend much deeper than merely the current child-care debate. Over the last fifty years rising levels of income and education, urban and suburban expansion, rapid technological development, and the ever increasing geographical mobility of American families have greatly affected American life styles and family life. In the late 1960s many Americans are awaking with a start to find that "family life" has come to be typified by an unpleasant picture. College-educated mothers are isolated in suburbia away from the intellectual, cultural, social, and governmental activity of the city; away from their husbands except for a few evening hours when the man they married is exhausted and requires rejuvenation for the next day's rat race; away from college and high school friends who live scattered around the entire nation; and away from their mothers who live perhaps a thousand miles off, isolated except for a few short visits a year from children and grandchildren.[69]

These young mothers can barely make friendships with their neighbors before their husbands' jobs require them to jump to another part of the nation. Between moves mothers' days are consumed by running children from school to dance class to baseball to friends. Breaks from taxiing little ones are filled with domestic chores made more complex by having to get shirts whiter than white while being limited by ecologists to low-phosphate soaps, and having to compete with television ads as well as neighbors for the shiniest floors and table tops. The typical husband leads an exciting life compared to that of his wife. Daily he goes off to the big city. But doing so entails a boring commute robbing each of his days of two to three hours. While in the city he struggles to remain subservient and polite to the boss despite subtle insults and finds that competition with the other men prevents real friendships. Back in the greenery of suburbia the kids too compete. Their competition, like father's, allows only superficial friendships and creates an alienating gulf between them and the world.

In Chapter 4 Safilios-Rothschild pointed out that this isolated familial existence is unique in the world's history. Where does it lead? The statistics are well known. One in four marriages break up. Children seek escape through drugs. Mothers turn to alcohol. Fathers flee from one unhappy marriage and soon find themselves in another.

Today more and more young people, joined by a few middle-aged divorced friends, are searching—sometimes singly, sometimes in couples or threes and fours—for an alternative to this life. Over 140 communes are listed by a local agency in Boston alone. Young couples, realizing that there is no logical ground for the sexual division of childrearing and other housework (except reproduction, which takes little time), are sharing domestic chores and childrearing equally, sometimes alternating domestic and work roles on a yearly basis, sometimes both working in jobs part-time and in the home part-time (see Chapter 17 by Elizabeth Hagen).[70] Infant- and child-development services greatly help parents by partially freeing them of child care, by giving them professional advice and opinions regarding their child, and by allowing them to feel that they are not alone in this important responsibility. Many couples who have experimented with shared roles are happy with them. Mother is relieved of the monotony of housework and of suburban isolation. Father is relieved of the burden of being the family's sole economic support. Mother can develop some of her occupational and professional interests. Father gains the joy of seeing his children grow and develop. He is no longer a stranger but a very important person to his offspring.

Much more familial experimentation will be needed before we, as a society, are likely to find a familial structure and life style that provide both human fulfillment and security in our age of rapid change and complexity. By socially condemning and legally outlawing experimentation and by granting governmental support solely to traditional nuclear families, our society could prevent all alternative forms of life from being proven more successful and lock itself permanently within the isolated nuclear family structure. It is to be hoped that we will not suffer this myopia.

Conclusion

The preschool years are nearly one-tenth of the average person's life. They are also the basis for an individual's development throughout the rest of life. Do we value these years of human life? Today we obviously value inequality, military aircraft, highways, and lunar landings more. As the richest nation in the world, we placed the first man on the moon while our infant mortality rate is higher than that of a dozen other nations. Thousands of children of working mothers are currently left to care for themselves or are cared for by a sibling who is still a child.[71] When we fully value children and spend our important dollars on them, we will not only provide society's offspring with quality child-development services, but also guarantee families adequate housing, good health care, recreation, and a decent, secure income, and offer each child a future toward which he or she may look forward.

NOTES

1. S. 2007, the Economic Opportunity Amendments of 1971, vetoed by President Nixon December 9, 1971.

2. The reader is referred to Chapter 5 for an analysis of how current congressional proposals treat these issues and goals. The issues of type of care (family day care or child-center care), the location of centers, and nonprofit or profit-making auspices for child care are discussed fully in Chapters 14 and 15 by Susan Stein and Janet Burton.

3. *New York Times*, December 10, 1971.

4. Nationally child-care costs are primarily related to the proportion of the early childhood population included in child-development programs. In terms of governmental budgets the cost of child care is also related to the proportion of care provided through the public as opposed to the private sector. Each of these factors will be discussed later under related issues.

5. U.S. Congress, Senate, Subcommittee on Employment, Manpower, and Poverty, and the Subcommittee on Children and Youth, *Hearings on S. 1512 to Amend the Economic Opportunity Act of 1964 to Provide for a Comprehensive Child-Development Program in the Department of Health, Education, and Welfare*, Part 2, 92nd Cong., 1st sess., May 25 and 26, 1971 (Washington, D.C.: U.S. Government Printing Office, 1971), p. 643. (Hereafter referred to as *Hearings on S. 1512*.) Cf. Mary Dublin Keyserling, *Windows on Day Care* (New York: National Council on Jewish Women, 1972).

6. Statement of Congresswoman Shirley Chisholm on preschool education and day care before the Select Subcommittee on Education of the House Committee on Education and Labor, March 4, 1970.

7. Sheila Cole, "The Search for the Truth about Day Care," *New York Times Magazine*, December 12, 1971, p. 84.

8. *Hearings on S. 1512*, p. 752.

9. Cf. S. M. Miller, "Criteria for Anti-Poverty Policies," *Poverty and Human Resources* 3 (1968).

10. Statement of John Niemeyer, president, Bank Street College of Education, *Hearings on S. 1512*, p. 158.

11. Judy Kleinberg, "Public Child Care: Our Hidden History," *The Second Wave* 1, no. 3 (1971):24.

12. John Bowlby, *Maternal Care and Mental Health* (Geneva: World Health Organization, 1952).

13. Cole, *op. cit.*, p. 86; Marion Howard, *Group Infant Care Programs: A Survey* (Washington, D. C.: Research Utilization and Information Sharing Project, George Washington University, February 1971); L. J. Yarrow, "Maternal Deprivation: Toward an Empirical and Conceptual Re-evaluation," *Psychological Bulletin* 58 (1961):459–490; Milton Willner, "Day Care: A Reassessment," *Child Welfare* 44, no. 3 (March 1965):127.

14. Cf. Westinghouse Learning Corporation and Westat Research, Inc., *Day Care Survey, 1970*, contract OEO BOO–5160, report of the Office of Economic Opportunity pursuant to April 1971, Table 4.13, p. 161.

15. U.S. Congress, Senate, Subcommittee on Children and Youth, *Hearings on Child Development Recommendations of the White House Conference on Children*, 92nd Cong. 1st sess., April 26 and 27, 1971 (Washington, D.C.: U.S. Government Printing Office, 1971), p. 751.

16. *Hearings on S. 1512*, p. 367.

17. Louise Gross and Phyllis MacEwan, "On Day Care," *Women: A Journal of Liberation* 1, no. 2 (Winter 1970):27–29.

18. "Nixon Must Veto Child Control Law," *Human Events*, October 9, 1971, p. 1.

19. Russell Kirk, "$40 Billion for Day Care?" *Milwaukee Journal*, November 29, 1971.

20. White House Conference on Children and Youth, 1970, *Report to the President* (Washington, D.C.: U.S. Government Printing Office, 1971), p. 285.

21. Cf. statement of Maurine McKinley, associate director, The Black Child Development Institute, *Hearings on S. 1512*, p. 368.

22. White House Conference on Children and Youth, *op. cit.*, p. 285.

23. The Task Force members did split over the responsibility parents should have in hiring and firing all staff. "Those who believed that parents should hire and fire all staff, had in mind an orderly process with checks and balances so that staff could defend themselves against any unfair attack. Those who believed that only the Director should hire and fire staff had in mind a Personnel Practices Committee from the Board to assist the Director so that a strong role for parents came out of the discussion in either case. Gwen G. Morgan, "Issues in Delivery of Day Care Services," talk given at Conference on Delivery of Day Care Services, Day Care and Child Development Council of America, Washington, D.C., July 25, 1971, p. 8; Alfred Kahn, "Report of Task Force on Delivery of Services to Developmental Child Care Forum," White House Conference on Children and Youth, December 1970, draft.

24. However, these minority group representatives, like many majority group liberals, do believe integration is needed to enrich the sterile homogeneous environment of middle-class suburban youngsters and to allow suburban children to learn to live and play with others of different backgrounds. Cf. Samuel Bowles, "Unequal Education and the Reproduction of the Social Division of Labor," in Richard Edwards, Michael Reich, and Thomas Weisskopf, eds., *The Capitalist System* (Englewood Cliffs, N.J.: Prentice-Hall, 1971); Statement of Cecilia Suarez, executive committee member, Mexican American System, *Hearings on S. 1512*, pp. 351–354.

25. Ruth Turner Perot, "Day Care and the Black Community: Myths and Realities," lecture in the Merrill-Palmer Institute 1971 Lecture Series, Detroit, April 27, 1971, p. 11.

26. Washington Research Project and the NAACP Legal Defense and Educational Fund, *Title I of ESEA: Is It Helping Poor Children?* 2nd ed. (Washington, D.C.: Washington Research Project, December 1969).

27. *New York Times*, December 9, 1971.

28. Sar A. Levitan, *The Great Society's Poor Law* (Baltimore: The Johns Hopkins University Press, 1969), p. 116; Daniel P. Moynihan, *Maximum Feasible Misunderstanding: Community Action in the War on Poverty* (New York: The Free Press, 1969); Robert A. Levine, *The Poor Ye Need Not Have With You: Lessons from the War on Poverty* (Cambridge, Mass.: MIT Press, 1970), pp. 55–56; Polly Greenberg, *The Devil Has Slippery Shoes: A Biased Biography of the Child Development Group of Mississippi* (New York: Macmillan, 1969), pp. 259–329.

29. White House Conference on Children and Youth, *op. cit.*, p. 282.

30. Patricia Gerald Bourne, Eliott A. Medrich, Louis Steadwell, and Donald Barr, *Day Care Nightmare* (Berkeley: Institute of Urban and Regional Development, University of California, February 1971) mimeo; Gwen G. Morgan, "Issues in Delivery of Day Care," pp. 1–5.

31. Here I am borrowing from Gwen Morgan's terminology. Gwen Morgan, "Evaluation of the 4-C Concept," Office of Planning and Program Coordination, The Commonwealth of Massachusetts, 1971, typed draft, p. 2.

32. Gwen Morgan, "Overview of Publicly Funded Categorical Programs for Young Children," Office of Planning and Program Coordination, The Commonwealth of Massachusetts, 1971 (mimeo) p. 7.

33. The federal 4-C (Community Coordinated Child Care) pilot programs point the way toward a coordinated child-care *system* designed to meet the total needs of children and their families. Cf. Day Care and Child Develop-

ment Council of America, Inc., *Community Coordinated Child Care: A Federal Partnership In Behalf of Children,* final Report submitted to the U.S. Department of Health, Education, and Welfare under the provisions of DHEW Contract No. OS–70–79 and OEO Contract No. 389–4518, Washington, D.C., December 31, 1970.

34. Mary Rowe, conversation, October 1971.

35. Letter from Stanley J. McFarland, assistant secretary, National Education Association to Hon. John Brademus, May 28, 1971, and "Preliminary Report of the Ad Hoc Joint Committee on the Preparation of Nursery and Kindergarten Teachers," both in U.S. Congress, House, Select Subcommittee on Education, *Hearings on H.R. 6748 and Related Bills to Provide a Comprehensive Child Development Program in the Department of Health, Education, and Welfare,* 92nd Cong., 1st sess., May 17, 21, and June 3, 1971 (Washington, D.C.: U.S. Government Printing Office, 1971), pp. 428, 446.

36. *Hearings on S. 1512,* pp. 369–370.

37. Statement of Vicki Lathom, *Hearings on S. 1512,* Part 3, pp. 752–753; E. Belle Evans, Beth Shub, Marlene Weinstein, *Day Care* (Boston: Beacon Press, 1971), p. 56.

38. Conversation, E. Belle Evans, October 1971.

39. S. M. Miller and Frank Riessman, *Social Class and Social Policy* (New York: Basic Books, 1968), chap. 5, "The Credentials Trap"; David Hapgood, *Diplomaism* (New York: Donald Brown, Inc., 1971).

40. Alice M. Rivlin, "A New Public Attention to Pre-School Child Development," *Washington Post,* December 1, 1971.

41. Westinghouse Learning Corporation and Westat Research, Inc., *op. cit.,* p. 64.

42. Given the low wages of child-care workers today, increased demand for child care may be expected to lead to work organization and a strong push for decent wages similar to that experienced with low paid health workers under Medicare. Edward Zigler, former director of the U.S. Office of Child Development, gave these workers the new title of "child-development associates" and has termed their occupation a "profession." Although these actions are pleasing, they hardly take the place of money when one is poor. Cf. Edward Zigler, "A New Child Care Profession: The Child Development Associate," *Young Children* 27, no. 2 (December 1971):71.

43. U.S. Senate *Hearings on Child Development Recommendations of the White House Conference on Children,* p. 10.

44. *Ibid.,* p. 87.

45. Where available, the Labor Department is to utilize child-care facilities developed by HEW on a priority basis; when the Secretary of Labor provides child-care services under his own jurisdiction, he must obtain the concurrence of HEW with regard to the policies to be used in administering the services.

46. Gilbert Y. Steiner, *The State of Welfare* (Washington, D.C.: The Brookings Institute, 1971), p. 51.

47. *Ibid.,* p. 74 (parenthetical material added); cf. Blanche Bernstein and Priscilla Giacchino, "Welfare and Income in New York City," Center for New York City Affairs of the New School for Social Research, study conducted for the New York State Council of Economic Advisers, August 8, 1971; Raymond Glazer, "Considerations Relevant to Report 'Welfare and Income in New York City,'" Research Department, Community Council of Greater New York, August 13, 1971.

48. David G. Gil, "Mothers' Wages—An Alternative Attack on Poverty," *Social Work Practice* (1969), p. 190; cf. Urie Bronfenbrenner and Jerome Bruner, "The President and the Children," *New York Times,* January 31, 1972, p. 41.

49. John Hanrahan, "Wealthy Get Tax Breaks, Subsidies on 'Farmland,'" *Washington Post,* March 1, 1971, p. 1, 13.

50. Cf. Irene Lurie, "The Distribution of Transfer Payments among Households," in the President's Income Maintenance Commission, *Technical Studies* (Washington, D.C.: U.S. Government Printing Office, 1970).

51. Jessie Bernard, "Labor Force Participation, Employment Status and Dependency among Mothers," in Samuel Klausner, *The Work Incentive Program:*

Making Adults Economically Independent, final report to the Manpower Administration, U.S. Department of Labor, under contract No. 51–4069–01, forthcoming, ch. 7. Cf. Mildred Rein and Barbara Wishnov, "Patterns of Work and Welfare in AFDC," *Welfare in Review* (November–December 1971).

52. For overall critiques of H.R. 1 see the Center of Social Welfare Policy and Law, Columbia University, "H.R. 1: The Opportunities for Families Program and Family Assistance Plan" (New York, June 1971); and Elizabeth Wickenden, *Notes on H.R. 1* (New York: National Assembly for Social Policy and Development, Inc., September 8, 1971, January 4, 1972). For a critique as well as a description of the historical development of the American welfare system, see Frances Fox Piven and Richard A. Cloward, *Regulating the Poor: The Functions of Public Welfare* (New York: Pantheon Books, 1971); Steiner, *op. cit.*; Bruno Stein, *On Relief: The Economics of Poverty and Public Welfare* (New York: Basic Books, 1971).

53. Therese Lansburgh, statement before the Finance Committee, U.S. Senate, August 27, 1970, contained in the Day Care and Child Development Council of America, Inc., "Legislative Summary No. VII. The Family Assistance Act of 1970" (Washington, D.C.: DCCDCA, 1970), p. 53.

54. Under this proposal 10 percent of all funds would be allocated to child-center care, 45 percent to in-home care, and 45 percent to family day care.

55. The Center on Social Welfare Policy and Law, Columbia University, "H.R. 1: The Social Security Amendments of 1971: A Critique" (New York, 1971), p. 1.

56. Cf. statement of Maurine McKinley, associate director, Black Child Development Institute, *Hearings on S. 1512*, p. 368.

57. Cf. S. M. Miller and Pamela Roby, *The Future of Inequality* (New York: Basic Books, 1970).

58. Twelve percent went to the second quintile, 17.4 percent to the middle, and 23.5 percent to the fourth quintile. U.S. Bureau of the Census, *Current Population Reports*, P–60, No. 80, October 4, 1971, p. 28.

59. *Ibid.*; Cf. Pamela Roby, "Inequality: A Trend Analysis," *Annals of the American Academy of Political and Social Science* 385 (September 1969):110–117.

60. Cf. Howard M. Wachtel, "Looking at Poverty from a Radical Perspective," *Review of Radical Political Economics* 3, no. 3 (Summer 1971); Barry Bluestone, "The Tripartite Economy: Labor Markets and the Working Poor," and Howard M. Wachtel, "The Impact of Labor Market Conditions on Hard-Core Unemployment," *Poverty and Human Resources* 5, no. 4 (July–August 1970); Peter M. Blau and Otis Dudley Duncan, *The American Occupational Structure* (New York: John Wiley and Sons, 1966); S. M. Miller and Pamela Roby, "Education and Redistribution: The Limits of a Strategy," *Integrated Education* 6, no. 5 (September 1968).

61. Mollie Orshansky, "The Shape of Poverty in 1966," *Social Security Bulletin* 31, no. 3 (March 1968):14.

62. Ranking families by wealth, the wealthiest 1 percent own 31 percent of total wealth and 61 percent of corporate stock. Frank Ackerman, Howard Birnbaum, James Wetzler, and Andrew Zimbalist, "Income Distribution in the United States," *Review of Radical Political Economics*, 3, no. 3 (Summer 1971):25, 38; U.S. Internal Revenue Service, *Statistics of Income, 1966: Individual Income Tax Returns* (Washington, D.C.: U.S. Government Printing Office, 1968) tables 7, 11, and 19; Dorothy S. Projector and Gertrude Weiss, *Survey of Financial Characteristics of Consumers* (Washington, D.C.: U.S. Federal Reserve System, 1966).

63. Economic studies indicate that the overall distribution of taxes (including federal, state, and local sales, property, and income taxes) is proportional to the distribution of income rather than progressive. Cf. Henry Aaron, "Income Taxes and Housing," *American Economic Review* 60, no. 5 (December 1970); Joseph A. Pechman, "The Rich, the Poor and the Taxes They Pay," *Public Interest*, no. 17 (Fall 1969); George Bishop, *Tax Burdens and Benefits of Government Expeditures by Income Class, 1961 and 1967* (New York: The

Tax Foundation, 1967); Roger A. Herriot and Herman P. Miller, "Who Paid the Taxes in 1968?" U.S. Bureau of the Census, 1971 (mimeo).

64. The work registration requirement is likely to also make the unemployment rate soar since unskilled jobs are already scarce.

65. President Nixon, "OEO Veto Message," *New York Times,* December 10, 1971.

66. *Congressional Record,* Vol. 117, no. 196, 92nd Cong. 1st sess. December 14, 1971.

67. *New York Times,* December 11, 1971.

68. Cf. *Washington Post,* December 12, 1971.

69. Cf. Barrington Moore, Jr., "Thoughts on the Future of the Family," in Frank Lindenfeld, ed., *Radical Perspectives on Social Problems* (New York: Macmillan, 1968); Philip Slater, *The Pursuit of Loneliness: American Culture at the Breaking Point* (Boston: Beacon Press, 1970); Annabelle Motz, "The Family as a Company of Players," *Trans-action* 2, no. 3 (March 1965):27–30; Pauline Bart, "Portnoy's Mother's Complaint," *Trans-action* 7, no. 13 (November 1970).

70. Cf. Jessie Bernard, *Women and the Public Interest* (Chicago: Aldine, 1971), pp. 264–279; S. M. Miller, Pamela Roby, and Daphne Joslin, "Social Issues of the Future," in Erwin O. Smigel, ed., *Handbook on the Study of Social Problems* (Chicago: Rand-McNally, 1971); S. M. Miller, "The Making of a Confused Middle-Class Husband," *Social Policy* 2, (July 1971): 33–39; Donald W. Ball, "The 'Family' As a Sociological Problem," *Social Problems* 19, no. 3 (Winter, 1972).

71. Statement of Mary Dublin Keyserling, *Hearings on S. 1512,* p. 643; Seth Low and Pearl G. Spindler, *Child Care Arrangements of Working Mothers in the United States,* Children's Bureau Publication, no. 461 (1968), p. 71.

6

ONE STEP FORWARD—
TWO STEPS BACK: CHILD CARE'S
LONG AMERICAN HISTORY

VIRGINIA KERR

Introduction

For over a century day care has struggled to come of age as an American institution, and for as many years it has not succeeded. Although child care is not new to this country, those with an eye for drama or a taste for the heroics of social reform will find little satisfaction from examining its history. Official society, which has maintained a vigorous interest in protecting the unborn child and in educating children beginning at age five, has, except in certain crisis periods, given the mother sole responsibility for the intervening years. Indeed, if modern campaigners for extended day-care services were to draw on the lessons of history, they might be tempted to pursue their goals by creating World War III, a depression, or some similar catastrophe that would lift the stigma from the working mother and release support for services to her children.

Day care has at times been described as a partnership of three professions: health, social work, and education. In practice it has grown up as a footnote to those professions. Resisted by public-school people because it might compete for already strained funds and energy; recommended with reluctance by social workers who saw it as an unfortunate alternative to "normal" family life; and regarded with no enthusiasm by the organized health professions who saw extended group-care services as a version of socialized medicine, day care has been left to the whims of the charitable impulse or the profit motive. To trace its history is to delve into the esoterica of old conference reports and statistics on unmet needs only to emerge with a handful of random notations and the hope that the next 100 years will be better than the last.

The Day Nursery since 1838

Day care was an invention of the nineteenth century—a response to the immigration that brought over five million families to the United States between 1815 and 1860 and to the industrialization that took women who needed to work away from the home and into the factory. Children of immigrant families without relatives or other social connections who might have facilitated arrangements for care were left to fend for themselves in locked apartments or on the streets. In a situation ripe for philanthropic intervention, wealthy women, service institutions, settlement houses, or private individuals organized day nurseries to provide for the care and protection of children of working mothers.

Although the origin of the idea of caring for young children in groups is somewhat obscure, most who have cared to comment on the subject agree that the French *crèche* was the model for the American day nursery. *Crèches*, designed to reduce the high death rates of infants whose mothers worked in factories, grew up in France in the early 1900s; according to one report were inspired by a *garderie* started in Vosges, France, in 1770, by a clergyman who noticed a young village woman caring for the children of mothers who were working in the fields.[1]

The first U.S. day nursery was opened in Boston in 1838 by Mrs. Joseph Hale to provide care for the children of seamen's wives and widows. In 1854 the Nurses and Children's Hospital in New York City opened its version of the day nursery to care for children of working mothers who had been patients. Two women from Troy, New York, visited the hospital nursery, liked the idea, and opened their own in 1858. Connected with the Troy center was a clinic that provided medical care to the community as well as to the nursery children. During the Civil War the children of women who worked in hospitals and factories in Philadelphia were served by a nursery that opened in 1863.[2] In 1893 a model day nursery set up at the World's Fair in Chicago cared for 10,000 children of visitors. By 1898 around 175 day nurseries were operating in various parts of the country, enough to warrant the creation of a National Federation of Day Nurseries, which hoped "to unite in one central body all day nurseries and to endeavor to secure the highest attainable standard of merit."[3]

As one might expect, the quality of these early nurseries depended very much on the imagination and energy of the director. Those that had the advantage of strong leadership attempted, in addition to providing clean, healthy places for children, to offer something of interest for them to do during the day and to solve, for other members of the family, problems that came to their attention. Other nurseries were at best custodial holding operations that focused on physical care and protection from environmental hazards.[4] Also, surrounding the service was an undercurrent of bad feeling that faulted the users of the day nursery for being in a position of need. Such feeling was probably imported from England where poverty was thought to be a result of bad character and idleness and where aid to the poor, when given, was given grudgingly. American day nurseries often stipulated admission requirements such as the following from a day nursery started in Baltimore in 1888: "Children who can be well cared for at home will not be received, nor will children of idle, unworthy parents." [5]

Because the majority of the early day nurseries served immigrant families, they hoped, in addition to providing care and protection, to help socialize the children. Jane Addams describes the evolution of the Hull House day nursery as follows: "It is now carried on by the United Charities of Chicago in a finely equipped building on our block, where the immigrant mothers are cared for as well as the children, and where they are taught the things which will make life in America more possible." [6] Such things included manners, eating habits, and hygiene. The concern for socialization was reflected in the recommendations of the 1905 National Conference on Day Nurseries, which, among other things, suggested that day nurseries provide separate toothbrushes for each child and encourage children to brush at least once during the day. In California, where the first public-school nurseries were opened after the passage of a compulsory school law in 1910 as a measure against the truancy of older siblings, the nursery programs were designed "to begin proper education and to Americanize foreign children." [7]

Day nurseries, although usually supported by parents' fees (around ten cents a day) and private subscriptions, did receive some support from the states. For example, Maryland in 1901 donated $3,000 to support two day nurseries in Baltimore.[8] In addition there was some pressure from city agencies to increase the number of day nurseries because immigrant parents, lacking child care, were giving up their children to the care of public institutions at great expense to the public. In 1899, in New York City, 15,000 children were turned over to orphan asylums at a cost of over half a million dollars, a practice that led social agencies to recommend day nurseries as a "more humane and less costly method." [9]

Day nurseries, however, were by no means coming into their own as an institution. Social workers criticized their low health standards and suggested that they jeopardized home life. The first White House Conference on Children and Youth, held in 1909, heralded home life as "the highest and finest product of civilization" and urged that children be cared for in their own homes whenever possible. The Conference recommended mothers' pensions as a substitute for day nursery care, and by 1913 twenty states had enacted laws authorizing financial assistance to indigent mothers, with payments ranging from two dollars a week to fifteen dollars a month for the first child.[10] It was hoped that the pensions would enable mothers to stay at home to care for their children, but, as with the modern version AFDC, the payments were not adequate to the purpose, mothers continued to work, and day nurseries struggled without benefit of public support.

In general the late nineteenth-century movements for social reform were silent on the subject of day care. Abuses in foster homes and orphan asylums occasioned the passage of some state laws that gave Boards of Charities responsibility for licensing child-care facilities. In New York City the Bureau of Child Hygiene in 1905 required a licensed physician to give a medical examination to every child cared for in a nursery, but there was no regular inspection of facilities by health officials and little was done to close nurseries that fell below standard. Most licensing staffs were small, unclear about their responsibility, and more concerned with state-subsidized institutions than with day nurseries.[11] The women's suffrage movement concentrated on winning the vote, on the passage of child-labor laws, and on improving conditions for women in factories and had little to say about day care. Charlotte Perkins Gilman, who published *The Home: Its Work and Influence* in 1903, seems to have been a lone advocate of extended group-care services. "Our homes are not planned or managed in the interests of little children" said Miss Gilman, "and the isolated homebound mother is in no way adequate to their proper rearing." Casting a precocious eye on the myths surrounding the home, she suggested that day nurseries had a better record of care for children and, with what now seems like an untoward optimism, she said that "that which no million separate families would give their millions of children, the state can give and does."[12]

The Nursery School

During World War I the Women's Committee of the Council of National Defense discussed the possibility of establishing day nurseries in manufacturing areas, but no action was taken, and day-care needs created by the war were met in centers operated on an ad hoc local basis. The care and education of young children, however, became a national concern during this period because of the popular belief that problems in early childhood had caused the physical or mental deficiencies that disqualified so many men for military service. In 1915 a group of faculty wives at the University of Chicago organized the first U.S. cooperative nursery school "to offer an opportunity for wholesome play for their children, to give the mothers certain hours of leisure from child care and to try the social venture of cooperation of mothers in child care."[13] In the 1920s experimental centers that emphasized research in child care and development were opened in New York, Detroit, Boston, and at the Universities of Iowa, Minnesota, and California. Nursery schools, influenced by the ideas of Friedrich Froebel and Maria Montessori in education and claiming as a model Robert Owen's infant school opened in 1816 in New Lanark, Scotland, caught on quickly in the United States. But unlike their English models, which served poor children of working parents, American nursery schools, if operated privately, served middle-class children whose parents were eager to give them an early educational experience or, if connected with universities, concentrated on research, not relief.[14]

Although nursery school professionals did take pains to distinguish their service from that of the day nursery, the nursery school program, with its emphasis on education and development, had a positive influence on the day nursery both through publication of research findings and through placement of nursery school teachers in the day nurseries. Because the nursery school movement stressed the advantage of some group experience for children outside the home and proved the point by providing environments superior to many homes, it may have made a gentle dent in the fears that surrounded the practice of separating the child from its mother for any part of the day.

In general, though, the day nursery was regarded with a kind of contempt by nursery school people, and the relationship between the two institutions was not always smooth. The Child Welfare League, founded in 1920 to promote better standards for child care, attracted nursery school professionals and through its publications solidified the definition of the nursery school as an educational service and the day nursery as a service for families with some sort of social pathology.[15] Ethel Beer, who has written several fine books on the day nursery, suggests that nursery school teachers were often fatigued by the day nurseries' length and ill-prepared to meet the physical needs of the infants and sometimes of their families. She contends that the nursery movement, while making a positive educational contribution, did much to distract support from day nurseries and to intimidate the organizations that might have pressed for their expansion.[16] Her contention is borne out by the Education and Training Task Force of the 1930 White House Conference, which reported that the nursery school had captured the imagination of the middle classes and established educational institutions, while "the day nursery is almost invariably private, divorced from both our official health and our educational program, receiving little or no support from public funds."[17] Following the trend begun in 1909, social-service agencies, to which the day nurseries had their strongest ties, preferred to make grants to mothers rather than to institutions caring for children. Day nurseries, according to the Task Force, were often not available when needed or not used when available. All in all, the prognosis for the day nursery was unfavorable.

The Depression—Work Relief Nurseries

Day care had its largest growth during the depression of the 1930s with the creation of nursery schools financed by the federal government under the Federal Emergency Relief Administration and later the Works Progress Administration (WPA).[18] In 1933, to create jobs for unemployed nursery school teachers, President Roosevelt authorized expenditures for nursery schools to care for "children of needy, unemployed families or neglected or underprivileged homes where preschool age children will benefit from the program offered." All personnel, including teachers, cooks, nurses, nutritionists, clerical workers, cooks, and janitors, were to come from the relief rolls. The program provided money for in-service and preservice training for staff, encouraged parent education, and did much to increase public consciousness of the value of preschool education. Although the schools had been created to meet a welfare need, funds were administered through state departments of education and local school boards, and the nurseries were identified primarily as an educational service. By 1937, 40,000 children were being cared for under the program, which is still considered by professionals to have provided excellent health and nutritional care as well as education.[19] The WPA nurseries served the dual purposes of providing employment and of relieving some of the conditions of the depression that affected children adversely. A testament to the strength of the nursery school movement, the program represented the first federal recognition that the education and guidance of young children was a responsibility warranting the appropriation of public funds.

In what might best be seen as a separate but related development during the depression, Title V of the Social Security Act passed in 1935 allowed for grants-in-aid for child-welfare services, including day care, and also authorized grants for research in day care. At the same time the act provided for aid to dependent children deprived of a father's support "to keep the young children with their mother in the home." In most states the Social Security funds were administered through departments of public welfare, a precedent of much greater consequence for the future identity of day care than the WPA.

World War II—The Lanham Years

The U.S. entry into World War II occasioned a dramatic shift in the public attitude toward the working mother and sparked a major federal investment in the care of her children. For example, in 1943 *Time* reported the resistance to industrial day care on the part of U.S. welfare workers and federal agencies, who resented "the widespread notion that women who stay home to care for children are slackers." [20] The unease with which other articles reported the need for day care was no doubt a justified response to the unhealthy arrangements made for children while mothers flocked to work. War industries spawned boom towns; the locked car became a surrogate baby-sitter; and in many places bands of children roamed the streets while their mothers worked. Day care was very much in order, but its advocacy seems to have represented a tremendous psychological strain for a nation whose only unambivalent venture in child care had been an employment program for adults and whose social workers had been trained to advise mothers to stay at home.

In an ambitious attempt to accommodate both the ideal and the reality, the U.S. War Manpower Commission declared in 1942 that "the first responsibility of women, in war as in peace, is to give suitable care in their own homes to their children" and at the same time cautioned employers to set up "no barriers to employment of women with children," to provide flexible hours, shifts designed to coordinate with family life, and child-care services. [21] The nation—albeit guiltily—had recognized its responsibility to provide care for the children of working mothers, or at least the necessity of offering such care in order to attract women to war industries and keep them on the job. [22]

The money for day care came from a piece of legislation passed by Congress without the trauma or search for rationale that we have come to expect from federal child-care discussions. Public Law 137, the Lanham or Community Facilities Act, was passed in June 1942 "to provide for the acquisition and equipment of public works made necessary by the defense program" and contained not a word about day care. In August 1942, as a kind of inspired afterthought, a ruling specified that child-care centers in war-impacted areas could be considered public works. During the Lanham period the federal government spent $51,922,977 (matched by $26,008,839 from the states) on 3,102 centers, which served a total of 600,000 children. Forty-seven states took advantage of the program, although most centers were located in California, Washington, and New York. The effort was impressive, but it has been estimated that the Lanham centers could not have served more than 40 percent of the children in need of care. [23]

A number of federal and state agencies were involved, not too happily, in the administration of Lanham funds. In July 1941 the Children's Bureau convened a conference on working mothers that stimulated the formation of statewide child-care committees in nineteen states. In July 1942 the President appropriated $400,000 for the Office of Defense Health and Welfare Services (ODHEW) to make day-care planning and coordination grants to states whose plans were approved by the Office of Education (OE) or the Children's Bureau. Those with an affinity for rational planning might see these activities as preparation for the administration of federal funds through coordinating state agencies, but the Lanham Act, which was administered by the Federal Works Agency, provided for direct grants to local communities.

Although Lanham guidelines suggested that funding should go to programs approved by state committees in areas whose need was certified by the Office of Education or the Children's Bureau, the lines of authority were not clear.[24] In many cases the Federal Works Administration (FWA) made its own arrangements with localities, without benefit of consultation with either the states or the other federal agencies. Overlapping authority for day care—real or imagined—did not stop in Washington. In some communities applications were delayed because welfare, education, and recreation interests could not settle on an amicable division of responsibility for the programs.[25] The situation—rivalry among federal agencies for control of the program, conflict between states and localities—gave birth to an issue that has yet to be resolved: how shall federal day-care funds be delivered? Through state agencies or directly to communities? And to what kind of agency within communities?

The issue was discussed in hearings on a bill introduced in 1943 that would have divested the Federal Works Agency of its administrative authority for Lanham and channeled all funds through state agencies. The bill's supporters (ODHEW, OE, the Children's Bureau, and many day-care interest groups) argued that FWA's practice of making direct grants to localities offered no incentive for states to contribute to the programs, that FWA had not set up an evaluative mechanism, and finally that those agencies best prepared to administer child-care funds had been by-passed. FWA stood on its record and opposed the bill, with strong support from Alpha Kappa Alpha, an organization of Negro women, which charged the other agencies and the states with a history of discrimination.[26] FWA retained control of the program.

The most one can say about the explicit goals of the Lanham programs is that they were set up to allow mothers to work. Because the many WPA centers that received Lanham funding to continue had strong ties with schools and nursery education, the FWA was probably influenced to favor programs that were sponsored by education authorities. Centers were generally located in communities (as opposed to factories), and against opposition to care for children under two, they admitted infants because mothers demanded the service. Lanham also helped fund centers in farm labor camps for children of migrant workers, many of which had been started in 1936 by the Farm Security Administration and turned over to the War Food Administration when the farm program was disbanded in 1943.

The war was a time for innovation. In Los Angeles Gale Manor Apartments turned its ground floor into a combination nursery and playground and accepted only working parents with children as tenants.[27] The Kaiser Shipbuilding Corporation in Portland, Oregon, opened two day-care centers located directly at the entrance to each shipyard.

Buildings and equipment were provided by the U.S. Maritime Commission, and operating costs came from a combination of Lanham money, the company, and parents' fees. The director, James L. Hymes, Jr., author of Chapter 2 of this volume, described the philosophy of care as one of "meeting needs," which, among other things, included shopping for mothers, making doctors' appointments, mending clothing, caring for children who had minor illnesses, and providing carry-out dinners at low cost to parents who worked long hours. The Kaiser centers were open twelve months a year, twenty-four hours a day. It was an expensive program against which, according to all reports, there were no complaints.[28] But when the war ended the shipyards closed, and the centers closed with them.

Day Care after the War and during the 1950s

In one congressional discussion of Lanham Act day care, Carl Hayden remarked that "it is entirely proper that the Federal Government should appropriate child care money because Congress declared war, child care is a war problem, support will cease with the end." [29] So it did. In September 1945 nine national organizations appealed to the President to forestall the closing of the Lanham centers, but they only managed to postpone the death date until February 28, 1946.

When federal support ceased the Lanham centers closed in all but one state, California, which had a tradition of housing day nurseries in public schools and a strong lobby influencing the state to assume the program. In 1946 California's legislature voted to continue support of the Lanham centers, making two-to-one matching funds available on a year-to-year basis to local boards of education for operation of the program. Although the California children's centers are tied to education, they have followed the welfare practice of requiring a means test for admission and of favoring children of single-parent families. In 1957 the state gave the centers permanence by removing the year-to-year funding provision.[30]

New York City also continued to support day care. During the war, with the WPA centers about to close and the city ineligible for Lanham funds, a Committee for the Wartime Care of Children formed to press the city and state to pay for day-care services. In 1942 the New York State Legislature approved the Moffet Act, which set up a tripartite financing plan for day care in municipalities that could not receive Lanham funds. Under this arrangement the cost of day care was divided among parents' fees, the state, and the city. In New York City, partly as a result of Mayor LaGuardia's negative attitude toward working mothers, day-care funds were administered through the Department of Public Welfare, whose employees were mandated to investigate each application for day care, to review with the mothers the advantages and disadvantages of going to work, and to weed out those who were not deemed to require assistance or to need to work. Around forty-eight centers were funded in 1947 with a capacity to serve 4,000 children. When Governor Dewey announced his decision to terminate state aid for day care, his home was picketed by women's groups, but he refused to see them and called them Communists. The state discontinued its funds, but New York City, in response to pressure from a broad coalition of service organizations and parents', women's, and community groups, took over the cost of the programs.[31]

The attitudes of Governor Dewey and Mayor LaGuardia were indicative of the postwar reaction to day care. Although the war experience had had the effect of bringing education and welfare authorities closer together and of stimulating communities to organize for service, the incentive to cooperate was based primarily on the expected reward in federal dollars. Central to the return to normalcy was propaganda encouraging mothers to stay at home. Magazine articles extolled the virtues of motherhood and popularized research done in the 1940s that described the damage done to children such as orphans who had been institutionalized.[32] No attempt was made to distinguish between such experiences and group care for children who stayed with their families. The little federal day-care money that was left was channeled through the welfare system for programs that served, inadequately, the very poor.

With the outbreak of the Korean War there was a brief flurry of interest in day care. As the headlines shouted mixed messages—"U.S. To Call Mothers into Defense Plants"; "Women War Workers Opposed"; "Mothers Beseige Day-Care Centers"—Congress passed the Defense Housing and Community Facilities and Services Act in 1951, which authorized loans or grants for community facilities, including day care. However, when funds were appropriated, use of the money for day care was specifically excluded.

Nevertheless, mothers continued to work. In 1950 there were 4.6 million employed mothers, as opposed to 1.5 million in 1940. Those women who did not qualify for publicly supported day care turned to the proprietary center or nursery school or made individual arrangements that varied from baby-sitters or care by a relative to, in extreme cases, use of the older sibling and the latchkey. Whatever effort there might have been during this period to promote increased public support of day care was stilled by the strong sentiment against the working mother and by community suspicion of programs that seemed to extend the powers of the state. Fear of communism touched all areas of life and—the U.S. war and depression programs notwithstanding—the child-care center was regarded in some parts as a Russian invention. Day care became a field for the creative statistician, as federal agencies and local day-care groups patiently surveyed and resurveyed existing day-care facilities and issued a number of reports that, if they varied in emphasis, always came to the same conclusion: not enough day care. In 1959 a survey showed that five times as many women were working as in 1940, with day-care facilities available for only 2.4 percent of their children.

The 1960s

The child-care recommendations of the 1960 White House Conference on Children and Youth were prophetic of the incoherence that was to characterize federal programs for children in the decade that followed. Of eight separate resolutions dealing with day care, two suggested that mothers of young children should stay at home unless forced to work; two suggested that both industry and government establish day-care facilities for children of working mothers; two advocated that the government provide supplementary preschool services especially through an expansion of nursery schools to serve families at all economic levels; one proposed family day care for children under three; and the final statement asked that every center have at least one person qualified in early childhood education.[33] The resolutions reflected the tradition of regarding the nursery school as a positive experience and day care as an unpleasant necessity and highlighted the ambivalence that accompanied attempts to merge the two services. In spite of the economic egalitarianism in the nursery school recommendation, nurseries were clearly conceived as the only suitable type of service for upper-middle-income families, while their relative, day care, was endorsed with caution.

Throughout the 1960s, federal spending for day care increased significantly but in a pattern calculated to reinforce an already segregated system of services—public day care for the poor, private nursery schools or child-care centers for the affluent, and potluck for those families who fell in neither category. The Social Security Act was amended in 1962 to provide funds for day care through state departments of welfare and again in 1967 to provide 75 percent federal funding for day care of children of past, present, and potential welfare recipients, as defined by the states. As part of the antipoverty program, Head Start was inaugurated in 1964 under the Economic Opportunity Act (EOA) to provide compensatory education for disadvantaged children, a kind of modern version of the day nurseries' program of socialization, which attempted to bring immigrant children into the mainstream of American life. The EOA also provided grants for day care of migrant workers' children and supportive services for manpower programs. In 1965 Title VII of the Housing and Urban Development Act provided financial and technical assistance for day-care centers; in 1966 the Model Cities Act included day-care projects as part of a city demonstration program. A relaxation of the official disapproval of group care for infants came in 1967, with a provision to channel Head Start money into a number of demonstration Parent and Child Centers (PCC) for children under three.[34] And under the 1967 Social Security amendments some money was provided for day care of children whose parents enrolled in the Work Incentive Program. Most of the federal programs, with the exception of Head Start, were designed to provide income maintenance for the poor; all of the federal programs were restricted, by virtue of strong eligibility requirements, to the poor and served, along with existing social prejudices, to segregate children as effectively as any of the more discriminatory public-school systems.

During the 1960s, however, early childhood education came into full flower as a goal for all children and a recommended component of all programs. Head Start, popular interpretations of the work of researchers in early childhood development, and the marketing genius of the educational hardware and software industry put "cognitive development" into common argot and created a nation obsessed with the IQ's of its children. In a sense the respectability of early childhood education helped legitimize demands for day care, which was redefined in 1969 by the Child Welfare League to include child development and education as well as care and protection.[35] Advocacy organizations for children also followed the trend. In 1967 the National Committee for the Day Care of Children, which was founded in New York in 1960 to press for an expansion of day-care services, moved to Washington, renamed itself the Day Care and Child Development Council of America, and set about the uncomfortable business of merging the philosophies of education and social work to form a coherent program for change. Parents' demands for more developmental day-care programs for children, however, resulted in no significant public programs and were met by small profit-making centers or the larger franchise day-care chains.

In the late 1960s, with day-care facilities available for only a fraction of the five million children whose mothers were at work, a fresh and positive note for day care came from the women's movement, which, unlike its predecessor in the early twentieth century, gave day care high priority in its demands for change. Feminists challenged the assumption that day care is a welfare service for children whose mothers have to work, stressing that child-care services should be available to all working mothers regardless of their reasons for seeking employment or their economic status. Women's movement activities, along with pressure for more early childhood services and dissatisfaction with the welfare system, have had the effect of broadening the concept of the day care or at least of placing discussion of the service in a positive context. Federal response, however, has not gone much beyond a series of comprehensive child-care bills that have been introduced but not passed, or passed but not signed by the President, with some regularity since 1967. Such bills envision a system of services for young children set up under an authority independent of the welfare system, serving a mixture of economic groups, and providing a full range of developmental services for children.

While the movement in support of a comprehensive or universal day-care system has been gaining support, problems remain that make such a goal rather fanciful. Only seven states have initiated universal kindergarten programs,[36] and in forty-four states day-care licensing and administration is carried out by welfare departments.[37] Day care, although its status has improved, still labors as an institution that grew up with the dual stigma of being a charitable service for distressed families whose mothers violated conventional mores by working away from the home. And, while the inclusion of education in the goals of day care has been a positive development, it has resulted in the area of licensing in rigid educational criteria based on credentials, not performance, which have so plagued the public-school system.

In addition, day care continues to suffer as an institution in search of a reliable professional constituency. One does not have to go far even today to find a social worker or an early childhood educator who will comment on the need for more and better day care, and at the same time deprecate the use of day care by women who do not have to work. Without such a constituency the success of efforts to lobby for expansion of day care at local, state, and federal levels is contingent on the ability of its advocates to effect working coalitions among professionals and agencies competing for control of programs and among community and social reform groups who often balk at any signs of compromise to their particular philosophies of care. In New York City and California local and state support of child care was realized after the war only because such broad coalitions were organized and functioned long enough to see the legislation through. In both places such coalitions have relied heavily on support from day care's natural constituent—the working mother.

NOTES

1. Ethel S. Beer, "The Day Nursery," *Social Science*, October, 1942; Beer, *Working Mothers and the Day Nursery* (New York: Whiteside, Inc., 1957).
2. Beer, *Working Mothers and the Day Nursery*, pp. 35–40; cf. Robert H. Bremner, ed., *Children and Youth in America: A Documentary History*, 3 vols. (Cambridge, Mass.: Harvard University Press, 1970).
3. National Society for the Study of Education, *Pre-School and Parental Education*, 28th Yearbook (Bloomington, Ill.: Public School Publishing Company, 1929), p. 15.
4. Beer, *Working Mothers and the Day Nursery*, pp. 15–30.
5. Health and Welfare Council of the Baltimore Area, *Day Care Needs in Maryland* (Baltimore: Health and Welfare Council, 1964), p. 56.
6. Jane Addams, *Twenty Years at Hull House* (New York: The MacMillan Company, 1910), p. 169.
7. National Society for the Study of Education, *op. cit.*, p. 90.

8. Health and Welfare Council of the Baltimore Area, *op. cit.*, p. 54.

9. National Society for the Study of Education, *op. cit.*, pp. 89–95.

10. *Ibid.*, pp. 27–30.

11. Shirley Chisholm, "National Day Care Program Introduced," *Congressional Record*, May 18, 1971, extension of remarks, p. E4549.

12. Charlotte Perkins Gilman, *The Home: Its Work and Influence* (New York: McClure, Phillips, and Co., 1903), pp. 329–338.

13. National Society for the Study of Education, *op. cit.*, p. 29.

14. Beer, *Working Mothers and the Day Nursery.*

15. Chisholm, *op. cit.*, pp. E4547–E4548.

16. Beer, "The Day Nursery," p. 27; Beer, "The Neglected Day Nursery," *The Journal of Educational Sociology*, May 1940.

17. White House Conference on Child Health and Protection, *Report from the Committee on the Infant and Pre-School Child*, Section III (Washington, D.C.: U.S. Government Printing Office, 1931), p. 10.

18. Chisholm, *op. cit.*, pp. E4549.

19. *Ibid.*

20. "Marvelous for Terry," *Time* 41 (1943).

21. "Employment in War Work of Women with Young Children," *Monthly Labor Review*, December 1942, p. 1184.

22. Cynics suggested that employers preferred to hire married women with children because such women could be dismissed more easily after the war. Women's Bureau, U.S. Department of Labor, "Employed Mothers and Child Care," Bulletin No. 246 (Washington, D.C.: U.S. Government Printing Office, 1953).

23. Children's Bureau, U.S. Department of Health, Education, and Welfare, *Spotlight on Day Care* (Washington, D.C.: U.S. Government Printing Office, 1966), p. 27.

24. "Whose Baby?" *Business Week*, March 20, 1943, pp. 32, 34, 37.

25. Kathryn Close, "Day Care Up to Now," *Survey Midmonthly* 79 (July 1943):194–197.

26. U.S. Congress, Senate, Committee on Education and Labor, *Hearings on S876, A Bill to Provide for the War Time Care and Protection of Children of Employed Mothers and on S1130, A Bill to Provide for Care of Children of Mothers Employed in War Areas in the United States and for Other Purposes*, 78th Cong., 1st sess. (Washington, D.C.: U.S Government Printing Office, 1943).

27. "Eight Hour Orphans," *Saturday Evening Post*, March–April 1943, pp. 24, 26.

28. Lois Meek Stolz, "The Nursery Comes to the Shipyard," *New York Times*, November 7, 1943, pp. 20–39.

29. *Ibid.*, U.S. Congress, Senate, Committee on Education and Labor, *op. cit.*

30. Elizabeth Prescott, Cynthia Milich, and Elizabeth Jones, *Group Day Care: A Study in Diversity* (Pasadena, Calif.: Pacific Oaks College, 1969), p. 26.

31. Chisholm, *op. cit.*, pp. E4549, E4551; cf. Cornelia Goldsmith, *The Story of Day Care in New York City* (Washington, D.C.: National Association for the Education of Young Children, 1972).

32. Cf. Nathan Maccoby, "The Communication of Child Rearing Advice to Parents," The *Merrill-Palmer Quarterly* 7 (1961):199–204; Clark Vincent, "Trends in Infant Care Ideas," *Child Development* 22, no. 3 (1951):199–209; Milton Yinger, "The Changing Family in a Changing Society," in *Family and Change*, ed. John Edwards (New York: Knopf, 1969).

33. *Proceedings of the Golden Anniversary White House Conference on Children and Youth*, March 27–April 2, 1960 (Washington, D.C.: U.S. Government Printing Office, 1960), recommendation numbers 358–365.

34. As of June 1, 1970 there were thirty-six PCC's in operation. Seven are administered by OEO and focus on research; the rest, under HEW administration, are primarily service-oriented. The programs vary in emphasis—one serves teen-age unmarried mothers and their children; another hopes to estab-

lish a day-care certification program for the mothers; others concentrate on improving cognitive development of infants. All Parent and Child Centers include parental involvement, planned activities for children, and comprehensive health care for children and their families. Alice M. Pieper, "Parent and Child Centers—Impetus, Implementation, In-Depth View," *Youth—Children*, December 1970, pp. 70–76.

35. Bettye M. Caldwell, "A Timid Giant Grows Bolder," *Saturday Review*, February 20, 1971, p. 48.

36. Marian Wright Edelman, "Statement Before the Joint Hearing of the Senate Subcommittees on Employment, Manpower, and Poverty and on Children and Youth," Washington Research Project, May 25, 1971, p. 8 (mimeo).

37. Day Care and Child Development Council of America, *Basic Facts about Licensing of Day Care* (Washington, D.C.: Day Care and Child Development Council of America, October 1970), *passim*.

7

UNIVERSITY DAY CARE

ADELE SIMMONS AND
ANTONIA HANDLER CHAYES

Over the past decade universities have become a major focus for the child-care movement. By 1971 many universities had considered the possibility of opening a day-care center, and a few had assisted in starting centers on or off campus. That day care became a major campus issue is not an accident, but rather a reflection of the special nature of the constituencies that advocate day care and the unusual decision-making process of most universities. The two have interacted to create a climate generally favorable to day care. In this chapter we will discuss the political process involved in bringing child care to universities, describe the establishment of day care at Tufts University, and then consider the special role of university day care in the child-care movement.[1]

Parties, Players, and Pressure Groups

Before discussing the university decision-making process, it is important to describe those groups that seem repeatedly to bring the most pressure to bear on university decision-makers with regard to child care. While university communities tend to be fragmented and include many interest groups with conflicting priorities, day care is one issue that has attracted the support of disparate groups representing the major political forces on campus. These advocates of day care include: women graduate students; women faculty members; faculty wives; undergraduate students —some of whom belong to political action organizations—and employees. These constituencies, sometimes working separately and sometimes together, have pressed university administrators to establish day-care centers. While they share day care as a common goal, their motives and their understanding of quality day care differ widely. The process of bringing child care to campus involves the reconciliation of these differences as much as it requires persuading the administration and faculty that child care is essential.

The first group to urge day care is likely to be women graduate students, supported by women faculty members and faculty wives.[2] For the most part these are women who have invested in graduate or professional training and want to work or continue their studies. They have spent many frustrating hours working out adequate child-care arrangements to free them for a few hours of studying and writing. As a result of consciousness-raising by the women's movement, they have become increasingly dissatisfied with makeshift child-care arrangements and makeshift careers. Because they have flexible work schedules, it is easier for them to meet regularly. They tend to be familiar with the university decision-making process and are able to develop effective strategies to bring day care to the attention of the university administrators. Finally, because they are both educators and mothers, they are confident that they can set standards for good child care, even though they may have no special training in child development.

The faculty and graduate student mothers who work for day care do not belong to one common political faction. Some have worked closely with radical students, often through the New University Conference (NUC), a national organization of radical faculty members and graduate students, which has been instrumental in the day-care movement on nearly every campus; others have close connections with the administration. Many have a legitimacy within the university that comes with faculty rank and are respected precisely because they have not been active in political events of the past four years. Their arguments are strong and widely supported. They argue that the university should be a model institution for equal opportunity, but that its present record is dismal. Women continue to suffer discrimination. Until day care is readily available, women cannot have equal opportunity in any job.[3] In fact beginning a day-care center is one of the easiest responses for a university pressured to improve the status of women. Hiring more women faculty members requires the cooperation of often intransigent department chairmen, particularly at a time when many universities are limiting faculty hiring. But a day-care center is a fringe activity that has no direct impact on academic departments, except for those that use it as a laboratory.

A second and more vocal group of day-care advocates are undergraduate students. With the exception of those students who are mothers, undergraduates who participate in day-care movements on campus usually belong to a radical political action group that works with radical faculty. More than a woman's issue, radicals see child care as a first step toward dramatic social change. They hope that through the day-care movement "a unified group of campus workers, college teachers, and students may emerge to carry on other battles."[4] They also see day care as a first step in the necessary breakdown and broadening of the nuclear family. For example, in 1970 the Women's Caucus of the New University Conference suggested that "through day care we can begin to break down the notion of the child as private property." [5] Similarly, a University of Indiana, Women's Liberation leaflet said that child care would give children "an opportunity to break out of the repressive family situation . . . to let children grow, to function with as much physical and emotional independence as they need." [6]

Because child care is part of a larger political and social program, they have clear views about the nature of university child care. It should be free and available to members of the community surrounding the university as well as to all members of the university community. It should teach values of particular importance to radical groups. A child-care center that emphasizes collective rather than individual goals, that places little emphasis on individual achievement, and that includes children of workers as well as children of the middle class can, such political groups argue, move society toward socialism.

Radicals are also the first to point out that day care provides the university with an opportunity to serve the community in which it is located. They point out that even though the community permits the university to function on a tax-free basis, the university provides the community with little in return. While it is unrealistic to expect universities to meet all the day-care needs of an urban area even with massive funding, the prospect of community involvement attracts a number of supporters to the side of day care.

Radical students have promoted day care through continual lobbying and publicity as much as by threat of disruption or actual disruption. However, the previous history of campus disturbances lends considerable effectiveness to the implied threat. In the few cases where disruption has occurred, it has usually taken the form of a sit-in (sometimes called a cry-in or a baby-in) in an administrative office or the establishment of a temporary center in space not approved by the university administration. For example, such disruptions have taken place at the University of Washington, Temple University, and the University of North Carolina. At Temple University the administration sought a court injunction to force parents and children out of a student study lounge that was being used as a center. [7]

Once radical students raise the issue of day care, other undergraduates become involved. Moderate students are attracted to day care as one of the few immediate and tangible issues available, less remote than the Vietnam War or government contracts for research, and one in which they can use their energies and campus-organizing abilities. On some campuses these students have contributed to the initial funding of a day-care program. At Tufts the Class of 1970 gave $5,000 to help cover start-up costs. The student government at the University of Indiana provided $1,500 for day care. Ten thousand dollars in student fees were allocated to day care at the University of Oregon. The student council at

California State in Hayward has voted to allocate seven dollars of each student's fees to the day-care center.

Black students, while supporters of day care, have not thus far participated actively. They tend to be more concerned with issues that affect Black undergraduates directly. Furthermore, suspicion between white radical political groups and Black students is such that once day care becomes an issue for the white radicals, Black students tend to stay away. Black ambivalence toward women's liberation also explains the coolness to day care as a university issue, though it is an important ghetto and welfare issue.[8] The perceived potential of Black student involvement in the day-care issue does, however, constitute a pressure on decision-makers, particularly when day care is related to attracting minority employees to campus, as described later.

Nonacademic staff is a third campus group that advocates the establishment of day-care facilities. Within this constituency the most active child-care supporters are Black employees, backed by those who are anxious to increase the number of Black employees. In response to pressure from the federal government, universities have developed affirmative action programs committing themselves to employing minorities.[9] Several universities, including Harvard, have faced probing investigations about their employment policies in respect to women. The University of Michigan has in fact had funds temporarily withheld by HEW until they developed an affirmative action program that eliminated the discrimination against women that investigations found. Day care can thus be linked to affirmative action, for it is viewed as one way to attract minorities to a campus in an urban area with sizable minority enclaves.[10]

Day care increases the pool of potential employees by freeing mothers of small children for work. In addition good child care at the place of work can compensate for the low wages that universities traditionally pay and for the long commute necessary to reach the university from the inner city. Finally, as long as education is linked with upward mobility, many parents, both minority and nonminority, are willing to make some sacrifices to ensure that their children have educational opportunities. University day care is presumed to be educative day care.[11]

Nonacademic employees, other than minority employees, have so far remained aloof from the day-care issue on many campuses. Many came to work before day care was available and have made other child-care arrangements or do not have small children.[12] Moreover, they are rarely included in the decision-making process and have little history of political action.[13] However, as unionization expands among university employees, day care is likely to become a subject for collective bargaining on campuses.[14] Some universities already contribute to employees' child-care costs in a limited manner. Radcliffe, for example, already contributes half tuition for a limited number of employees' children to attend the Gymnasium Center. Through a grant from the Mellon Foundation, MIT is now able to pay approximately two-thirds of the difference between what an employee can afford and the $42.50 weekly tuition at the KLH center for fifteen employees.[15] Yale and Syracuse also have subsidized programs.[16] It is easy to anticipate that the nonacademic staff as a whole will become a much more forceful pressure group in the future.

The major pressure groups described above can converge to bring substantial pressure on a university administration, which is ultimately responsible for the establishment of child care. For the administration, negotiations with these constituencies, which include moderate faculty, students, and employees along with the most radical groups, tend to be easier and less hostile than those involving radical groups alone. At the same time the administration finds it politically easier to meet the demands of such a broad spectrum than to meet the demands of SDS and Progressive Labor, groups against whom administrators are being encouraged to take a strong position.

It is still unclear whether it is politically more effective to form a coalition for day care composed of as many groups as possible or to have each group present its own proposal to the administration. While all interested parties must ultimately agree on guidelines for day care, multiple proposals often convey the strength of the various day-care constituencies. In either case, the force these pressure groups bring to bear does not alone explain why, in the face of serious financial difficulties, universities have taken the risks associated with starting a day-care center. In part the explanation lies in the university decision-making process, which reinforces and encourages the establishment of day care.

University Decision-Making

The classic organization is hierarchically arranged, and while decisions often diverge from the organizational model, this structure leads to certain expectations about the flow of authority.[17] In contrast university organization is not hierarchical. It has two distinct focal points: (1) the administration, ordinarily hierarchically organized, and (2) the faculty, organized by department, often working through committees or through a representative body such as a Senate.[18] A number of factors are critical to the balance of decision-making power in any university: among them the effectiveness and personalities of the president and his staff; the quality of the deans; the unity of the faculty; and the wealth or poverty of the endowment, or the relations with the legislature in the case of public institutions. The division of labor and the interaction between the two focal points defines the decision-making process somewhat differently in each institution. Nevertheless, a set of expectations has evolved about this organizational model based on the jurisdiction or province of each focal point.

Day care represents an intriguing case of university decision-making because it exposes the ambiguities about jurisdiction and the allocation of responsibility. It would seem that the lack of clearly defined boundaries has permitted the concentrated pressure of day-care constituencies to act as a wedge more powerful than the forces that could be marshaled to resist it. Just why this should be the case is worthy of careful analysis.

In theory *faculties*, working either collectively or representatively, are responsible for developing the curriculum and the overall educational policies and priorities of the institution. The *administration*, in addition to being charged with the day-to-day operation of the university—its housekeeping and service requirements—is responsible for financial and long-range planning and the implementation and accommodation of the educational policies established by the faculty or the faculty in concert with the deans. These areas of jurisdiction are never quite firm. The presidential power of the purse and the residual power of the trustees can alter the balance in establishing educational priorities. At the same time the policy of making university financial records public has increased faculty potential in long-range planning. They are now in a position to demand justification of programs in terms of establishing priorities among competing ideas.

Day care as an issue does not lie clearly within the province of the faculty or the administration, for it is both an educational program and a service to university employees. Ironically each has demanded justification in terms of the other's province. The faculty see day care primarily as a service, not as an educational program that intrudes upon an already strained instructional budget. As the Tufts case suggests, only child-study or child-development departments are likely to be interested in the training and research potential of a center. Even for these departments day care is likely to represent a threatening departure from the traditional laboratory preschool of the nursery school movement. At California State, Hayward, the child-development department has virtually ignored the day-care center. This pattern is not unusual.

For the faculty as a whole, the service appeal of day care is greater than the educational appeal, particularly when presented as a benefit for women, minority groups, and the local nonuniversity community. As pointed out, day care is a less painful adjustment to the requirement of equal educational opportunity than a rapid shift in sex ratios of faculty members.

The administration sees the importance of the service aspect of day care, but it also understands the financial implications of day care were it to become a subject for collective bargaining. The administration chooses, therefore, to demand educational justification in curricular and research terms, perhaps in part to assure continuing faculty effort to raise research funds to support the enterprise once it is started and in part to ensure faculty responsibility for a possibly costly addition to the university. In any event the administration seeks proof that the day-care center will add a new dimension for the undergraduate and graduate program, opening up opportunities for new courses, sponsored research, and practice-teaching experiences.

As both groups seek justification in terms of the other's province, so both the faculty and administration tries uneasily to shift the responsibility for making the decision onto the other. Both groups are loathe to oppose day care and be identified as the group responsible for blocking a program as sensible as day care. Thus both faculty and administration find it easier to give tacit support rather than active opposition to a day-care proposal. Unlike many current issues—Black studies, open admissions, or student representation on faculty committees—day care disrupts neither the educational program nor the administrative procedure. It appears to leave intact a carefully balanced decision-making process to which both the faculty and the administration are committed. Thus it appears far less threatening than many other issues with which the university decision-makers have had to grapple.

While both groups talk about the financial implications of day care, they are reassured by some of the pressure groups who explain that the center can be closed the year following a deficit. If this is not politically realistic, it is an understandable relief for administrators to be able to acquiesce in an issue that is relatively small, concrete, capable of execution, supported by a variety of politically important pressure groups, and opposed by no one.

Tufts·University: A Case Study

We have described the advocates and major parties involved in establishing child care. We have further attempted to outline briefly the university decision-making process and sketch the factors in it that foster day care. Now we will explore in detail the politics of day care at one university. Only when a number of case studies such as this are available will it be possible to take the hypotheses we have presented and develop a theory of university decision-making in regard to day care and similar issues.

The parties involved in bringing day care to Tufts University coincide, for the most part, with those described in the first section of this chapter. Specifically the major groups were: the office of the Jackson College dean, representing the interests of women; [19] the Office of Minority Affairs and the Afro-American Society; faculty committees, primarily the Educational Politics Committee; the New University Conference; various student groups, including SDS; a new dean of the Faculty of Arts and Sciences; and the provost. The process was complex and involved the breaking down of hostility toward the concept of day care as well as the establishment of a minimum of trust among those who had opposed each other in the past years of campus conflict. (In 1969 Black students led students and faculty in a confrontation with the administration about the lack of minority workers employed in the construction of a new dormitory.)

Child care first became an issue in the fall of 1968. A new dean of Jackson College (the women's college at Tufts), appointed to provide leadership in women's education, proposed a three-part program of continuing education, day care, and the study of family life. The dean turned to the Department of Child Study for support. Since a part-time nursery laboratory school—Eliot-Pearson—was affiliated with the department, expanding the department to include day care seemed a logical step. Most of the department members were willing to cooperate, but they had many questions about a program emanating from an office that had no experience with child development and had never taken leadership in academic areas before. They were concerned about the possible undermining or weakening of a successful traditional nursery school program, particularly in terms of financial and academic commitment as well as in terms of time and energy. The department was reluctant to take the risks that active involvement in day care required when no financial support seemed readily available. Department members helped the Jackson dean in discussing and contributing to proposals but did not provide the leadership.

The faculty meanwhile approved the proposal for continuing education and day care, subject to the availability of funds—a technique that faculties increasingly employ to indicate their approval for new programs without committing themselves to their actualization. By the spring of 1969, after nine months of proposal writing and grant seeking, no funds had been found and no visible progress had been made toward the establishment of a day-care center at Tufts.

Three significant changes in the spring of 1970 forced the university to look again at day care: (1) a sudden increase in the constituencies and numbers of the child-care advocates; (2) the opening of a Head Start program at Tufts; (3) the availability of outside funds.

In 1970 women's groups and radicals, through regional and sometimes national organizations, had begun to pinpoint day care as a priority. These groups had widespread affiliation on campuses among faculty and students. Their members were familiar with university politics. Since universities already had a reputation for responding to demands for new programs and radical change, they were a logical place to bring a new issue: day care. As on other campuses the first radical group to advocate day care at Tufts was the New University Conference. Composed of faculty and graduate students committed to changing traditional patterns of education, the Tufts NUC chapter began to work actively for day care in February 1970. The national NUC office had established definite criteria. University day care must be parent controlled and open to all members of the university community. It should be subsidized to bring it within reach of all workers. The program should convey attitudes toward "work, race, sex (including male and female roles), initiative and cooperation" consistent with those of radical groups.[20] Teachers at the center should be both male and female and should include minorities. Center hours should cover every employee's work schedule.

In articulating these national priorities, the Tufts NUC chapter considered itself in conflict with the Department of Child Study. It viewed the department's Eliot-Pearson as a school of middle-class values and sexual stereotyping. NUC rejected the notion of a laboratory center, as Eliot-Pearson was believed to be, where children's needs may be subordinate to research and training goals. Children should grow and develop and not be subjects for endless tests and experiments. Some of NUC's judgments about the child-study department and Eliot-Pearson reflected NUC's unfamiliarity with both. At the same time the department could not allay NUC's fears. Both groups failed to recognize that agreement between them was crucial. The child-study department needed NUC's political organization and commitment to sell child care. NUC needed the department's legitimacy to persuade the administration and faculty to accept day care; it also needed the department's experience and expertise in the field of child development. In any case NUC's initiative made child care a campus-wide issue. It led both the Jackson dean's office and the child-study department to renew their interest and it mobilized support from the newly created Office of Minority Affairs,[21] as well as SDS and other student groups.

The second impetus for child care at Tufts came unexpectedly in the spring of 1970. Because of a fire, a Somerville, Massachusetts, Head Start group was forced to find a new site. A Tufts administrator concerned chiefly with good community relations arranged for it to move into a rarely used lounge in a men's dormitory on an emergency basis. The Head Start program flourished and demonstrated that day care could operate on campus without elaborate preparation and equipment and without involvement of the child-study department, as NUC had hoped. Moreover, Head Start proved that the university could find space for day care.

Some of the child-care advocates suggested keeping Head Start on campus and adding a center with a group of equal size to create two classes. Each would include Head Start and university children. This unusual model might attract outside funding and at a minimum would assure regular Head Start funds. Such a center would have a broad social, economic, and racial mix and could serve to test the hypotheses of the Coleman report about the value of such mixed groupings of children.[22] This possibility, plus growing national interest in day care, led to a new focus for the potential Tufts center. Rather than designing elaborate and expensive teaching models, students and faculty would work toward refining and developing a model for low-cost quality day care. University centers would only have a wider relevance if their model could be used outside the university community. Head Start subsequently withdrew from discussions with Tufts because Head Start officials were concerned that a cooperative Tufts-Head Start Center would be significantly different from other centers. They were also concerned about the issue of control. However, the focus on low-cost quality care remained.

Outside funding was the third factor that contributed to the university's decision to bring day care to Tufts. The recent establishment of profit-making franchise centers forced a number of people to view with alarm the sudden growth of day care.[23] The New World Foundation, attracted by the Tufts proposal to establish higher educational standards in a relatively low-cost center that provided for university-community cooperation, and on the urging of the Jackson dean, gave the university a grant of $10,000. These funds were supplemented by a $5,000 gift that the graduating senior class voted to give the center. Money to start a center was at hand, provided that the university could find renovated space.

Fearful that once the start-up funds were used up the center would operate at a deficit and the university would be forced to close it, the administration remained reluctant to make a commitment to day care. Examples of other Boston area centers that had been raising tuition to meet costs did not help. The KLH Center had just raised tuition to a prohibitive $42.50 per week. In addition to its concern about finances, the administration, and the provost in particular, expressed concern about the educational component of the day-care program. The provost was firm in his belief that day care should be closely linked to undergraduate education and that the faculty needed to begin to assume responsibility for costly innovation. Thus he insisted that the issue be again considered by the major faculty committee, the Educational Policies Committee (EPC). Since the EPC did not meet over the summer, some viewed this as a delaying tactic. However, others recognized the political advantage of beginning the center with a clear indication of faculty approval and support.

Once the route to making a final decision about child care was clarified, child-care supporters began to focus on the nature of child care at Tufts. SDS and NUC came into conflict with the Department of Child Study and the administration over the issues of control and cost.

First, the provost's insistence that the center be directed by the Department of Child Study made NUC uneasy. They suspected that the provost's insistence upon tying day care to the department reflected as much his desire to find a legitimate means of excluding NUC from the governing of the center as his desire to relate child care to child study. Department control, a prerequisite for the educational program, could not be reconciled with parental control, a fundamental concern of all radical groups. In the end NUC and the Department of Child Study agreed to the principle of a governing board composed of a majority of parents, on the understanding that departmental approval was required for staff appointments.

Second, the radical groups insisted upon university subsidies to reduce tuition. Other day-care advocates, the Department of Child Study and the Jackson dean's office in particular, felt that it was futile to attempt to persuade the university to supply more than space, maintenance, and renovation and were skeptical that the university would even provide that much. The radical groups eventually split on the issue of funding, as well as on the interpretation of parental control. While NUC advocated parental control, organized meetings of potential parents, and then accepted the wishes of parents, SDS advocated parental control but then was confused by and unwilling to accept decisions by the potential parents that did not coincide with their views.

There seemed to be as much conflict as agreement among the day-care groups when the Educational Policies Committee met in the autumn of 1970 and assigned the day-care question to a subcommittee chaired by the new dean of the faculty, a Black psychologist who was known to be favorably disposed to day care. The appointment of the dean of the faculty as a committee chairman together with the appointment of representatives of all groups that had actively worked for day care suggested that the subcommittee would make a favorable report. The subcommittee worked hard, reworking old budgets to further reduce costs, obtaining what was taken to be a definite commitment on the part of the university to provide space and the $9,000 necessary to renovate that space, developing a model governing board similar to that worked out by NUC, the Department of Child Study, and the Jackson dean, and approving the principle of a sliding scale tuition with a budget that required an average weekly tuition of $17.50 per week with the help of the available outside funds.

In October 1970 the Subcommittee of the Educational Policies Committee reported to its parent committee. About twenty SDS members came to the meeting to demand free day care. They stated vociferously that students would oppose day care on campus unless it was free. The financial vice-president attacked day care from the other side by arguing that the university could not afford to have any day care. He astonished many in the room by stating that the university lacked funds for renovation at that time, a statement directly contrary to the earlier assurances of the dean of the faculty. The subcommittee, with its carefully prepared proposal, was caught in the familiar liberal position between the radicals and the conservatives. The intervention of SDS alienated the faculty committee and nearly led to the defeat of child care. The contradictions about finances heightened the tension and confusion. The momentum seemed lost. The issue was tabled.

The committee met again in November. An elaborate agenda to provide for full discussion of day care had been prepared, and individual members of the EPC had been approached by proponents in advance. While SDS was invited to attend the meeting, its members could only speak after members of the committee had spoken. However, their presence was intimidating in view of their previous performance. The meeting was opened on a new note. Formal statements were made by the chairmen of the Equal Educational Opportunity and Equal Employment Opportunity committees. Both emphasized that day care was essential to an increase in the number of minority employees at all levels of the university. The impact of their statements was considerable. The arguments and inconsistencies of prior meetings were not raised again. The EPC adopted the day-care proposal with no further discussion, clearing the way for renovation, hiring of staff, and the opening of the center.

The two-year period that was involved in Tufts' developing a plan for a center and then obtaining the necessary faculty and administrative support is not unusual. Different constituencies worked sometimes in cooperation, sometimes in conflict toward a common goal. Each applied pressure and appealed to different concerns of the decision-makers. Although the present center does not fulfill the ideal of all groups—it is not free, as the radical groups requested, nor is it under close control by the child-study department—it does represent an operational compromise.

The political pressure of the day-care constituencies was particularly difficult for the administration to resist once the day-care advocates had prepared and agreed upon a budget that, for at least twelve months, required nothing except space and maintenance from the university. Although future funding was in doubt, it would have been awkward for the university to reject $15,000 in grants,[24] especially ones that had the prestige of the New World Foundation and the sentiment of the senior class gift.

The positive commitment of some and the apathy of others made it possible to push through a concept that had no enemies. Child care as an educational and political issue threatened no major interests nor the *modus operandi* of the university. For nominal cost it added a new dimension to the curriculum. Even the space that the center was allotted was vacant at the time. Under these circumstances who in the university could speak against child care once the administration was willing to assume the temporary financial risk?

Clearly at every university and in every community the establishment and continuity of day care is a dynamic political process. An awareness of the politics of day care can help groups shortcut the process and anticipate the frustrations, but it cannot eliminate them.

The University's Special Contribution to Day Care

Tufts is one of several universities that has made a well-publicized move to establish a child-care center. Many more universities are likely to follow. As government financing becomes available, resisting the pressures for day care will become harder. Each new university child-care center will increase the pressures in those universities without them. Universities tend to make decisions like lemmings, and while this tendency has been generally deplored,[25] it is welcomed in child care. In any event there is little evidence to suggest this tendency will end.

While these increased opportunities for child care on university campuses help meet national day-care needs, the universities' special responsibility toward the national day-care movement is of greater significance. Universities have the capacity to develop models of low-cost, high-quality day care; to offer continuing research on early childhood education specifically as it relates to day care; to provide specialized services to neighboring day-care centers; and to train staff for day-care programs.

Whether quality day care is available to all children whose parents want and need it is, to a considerable extent, a function of costs. Universities have the resources and the freedom to experiment with alternatives to reduce costs and improve quality. As educational institutions, many are exempt from state licensing and local zoning regulations, which often serve to increase cost without improving quality. Universities have experts available to improve equipment design and administrative management techniques as well as educational programs. Most important, though, is its competence in applying new findings and knowledge about child development to an active program of day care. University centers are continually working to improve their low-cost quality model, and they should be encouraged to share this information with local community groups. With the increase in franchise centers notably uninterested in educative child care, universities have a special obligation to provide alternative standards.

Universities can actively work to improve day care on a national and local basis through research, community programs, and training. While a number of universities have faculty members and graduate students engaged in research on child development, few university programs explore the relations between a day-care experience and child development. Existing information about age groupings, maternal deprivation, and institutional intervention needs to be expanded.[26] Infant care is just beginning to be explored. Longitudinal research on the effects of child-care arrangements on adult and adolescent lives is presently lacking.[27] Recently the University of North Carolina and Washington University in St. Louis have developed programs to study the effect of active intervention upon the intellectual and emotional development of children.[28]

Until more conclusive evidence about the effects of child care is available in a form that can be easily understood by the general public, latent hostility to child care will remain strong.

While the need to explore further the relations between child-care centers and child development is obvious, less obvious is the need to develop techniques for evaluating child-care programs. What do we mean by quality child care? Even before we can describe the features of quality care, we need to understand better what quality means and how to adapt this definition to different environments. Present practices of measuring reading ability or IQ are obviously inadequate even for older children, and observational techniques remain crude. If quality child care can be defined, some sort of cost-benefit analysis must be applied to child care.[29] Universities have the resources for providing leadership in developing methodologies for evaluation. They also have the resources for broadening the base of day-care-related research. Widespread and effective child care requires a great deal of information, only a portion of which is directly related to child development. Day-care research has so far been the province of child psychologists, but questions that are now raised require the analysis of many other disciplines. Issues of health, government, parental participation, financing, and transportation all require further exploration.

The relationship between university day-care programs and the surrounding community is of great concern to both groups. Even if the federal government provides massive support for day care, universities cannot provide centers for the vast numbers of children needing day care. However, universities can offer the technical help that their surrounding communities will need to begin and operate a high-quality center. University specialists can advise neighboring centers on budgeting, designing of space, and the development of curriculum. A university-based community consulting service, if designed appropriately, could also be useful. In addition the university can offer courses on day care for the staff of neighboring centers and for parents whose children attend any center in the community.

Traditionally educators have demonstrated only a minimal concern about day care. As Virginia Kerr points out in Chapter 10, education and day care have long been separated conceptually. Educators have thought little about developing quality day-care programs or training personnel for day-care centers.[30] The Comprehensive Child-Development Bill presently before Congress asks for forty million dollars for training, yet few of our universities are equipped to provide such training. Curricular and often faculty changes will be required. A broader student body can be recruited, particularly if the university is willing to offer training to paraprofessionals as part of an Associate of Arts program.[31]

For many universities a strong academic program focusing on day care will coincide with changes in graduate and undergraduate education already under way. A plan of study in day care will encourage the student to draw upon a variety of disciplines in solving a broad spectrum of problems, thus enabling the student to acquire a liberal education through a problem orientation. At the same time day-care training offers opportunities for developing work-study programs and for expanding existing opportunities that provide academic credit for work experience. Thus while the day-care movement needs universities to train teachers, conduct research, and maintain educational standards, the universities can use day care as a focus for educational innovation.

Conclusion

Beginning day care in any community is a political process. While the politics of university day care are unique in some respects, there are lessons for nonuniversity groups interested in day care. The sources of support, both political and financial, are often the same. The issues about the nature of the center, parental control, a sliding scale tuition, and curriculum must also be considered by nonuniversity centers. There is a cohesion within the university community that often forces the resolution of conflict over these issues. It is clearly in the best interests of groups involved to succeed in establishing some kind of center rather than none at all. Furthermore, many of the people involved have had experience in resolving conflicts among themselves over other university issues. The compromises that result from conflicts at universities can in many cases serve as models for nonuniversity centers.

As university centers grow, the feasibility of establishing day-care centers at other institutions will become more evident. However, only massive federal funding can make day care universally available in the United States. In anticipation of such federal commitment, universities can now develop models for the new day-care centers with a solid educational component. An intensive and immediate investment on the part of universities in day care—research, community involvement, and training—can have a lasting impact on the burgeoning day-care movement. Universal day care should not be custodial care. Universities as institutions can help ensure that this does not happen.

NOTES

1. Our description of the process of establishing day care on campus is based on observation of a number of universities, including Boston University, Brandeis University, Brown University, University of California, University of Chicago, Harvard University, University of Indiana, Massachusetts Institute of Technology, Oregon

State, University of Pennsylvania, Radcliffe College, Temple University, Tufts University, University of Washington, and Washington University at St. Louis. It does not represent a systematic study of a statistically valid sample. However, because of the number of universities presently considering day care, we feel it is useful to offer a tentative hypothesis at this time and hope for a more complete study later. Cf. *LCCCCHE Newsletter,* Liaison Committee for Child Care Centers in Higher Education, 304 Eshelman Hall, University of California, Berkeley, California, 94720.

2. See, for example, Eleanor McLaughlin, "A Pilot Day-Care Facility, a Proposal for Wellesley College," and New Opportunities for Women Committee, "A Day-Care Proposal for Harvard" (Cambridge, 1970). Radcliffe alumnae were also actively involved in developing day-care proposals for Harvard and Radcliffe. Women's groups also raised the day-care issue at the University of Wisconsin, the University of Pennsylvania, and Brown University.

3. Nearly all university reports on the status of women mention child care as an essential prerequisite for equal opportunity for women. See, for example, Bernice Neugarten, *Women in the University of Chicago* (Chicago, 1970), p. 18; Janet Abu-Lughod, "A Proposal Concerning Current Higher Education at Northwestern University," 1970; (mimeo); Elizabeth Colson and Elizabeth Scoll, "Report of the Subcommittee on the Status of Academic Women on the Berkeley Campus," 1970, p. 8, Caroline Bynum and Michael Walzer, "Report of the Committee on the Status of Women in the Faculty of Arts and Sciences, Harvard University," May 1971.

4. Norman Daniels, "Why N.U.C. Should Fight for Child Care" (Medford, Mass.: Tufts University, Spring 1970).

5. NUC Women's Caucus, "Politics of Day Care," February 1970.

6. NUC Women's Caucus, "Politics of Day Care." Elsewhere Richard Lichtman has observed:

> The desire to fight for day care at Harvard and suspicion and militancy toward the administration on this issue cannot be understood purely in terms of Harvard's record of not having day care. We choose to fight for it both because it is something lots of us need, and because it is an example of the way Harvard and all major capitalist institutions re-enforce the oppressive features of the nuclear family ("Ideological Functions of the University," *International Socialist Journal,* no. 24, December 1967.

7. "Temple University Day-Care Fight," *Action for Children* 1, no. 2 (November 1970).

8. Toni Morrison, "What Black Woman Thinks about Women's Lib," *New York Times Magazine,* August 22, 1971.

9. The federal government has established guidelines for minority and female employment under Executive Order 11246 as amended by Executive Order 11375 3 C.F.R. 402 (1970). The executive order requires that federal contractors "shall not discriminate against any employee or applicant for employment because of race, color, religion, sex or national origin" and requires further that the contractor "take affirmative action to ensure that applicants are employed, and that employees are treated during employment without regard to their race, color, religion, sex or national origin."

10. Loretta Stokes, "Memorandum on the Need for Day-Care Services among Employees," Harvard University, April 1, 1971.

11. Over half of the minority female employees who joined the Tufts payroll since the day-care center began use the center.

12. At Harvard, for example, two-thirds of the nonadministrative staff is female, but these women are usually single, newlyweds, or over forty. New Opportunities for Women Committee, *Op cit.,* p. 6.

13. An important exception to this is the nonacademic staff at MIT. This group has been outspoken in its demand for day care and successfully contributed to pressures that led to the employment of a full-time staff member in the planning office to explore day-care possibilities at MIT. MIT Planning Office, "Day-Care Services: Research and Demonstration Program," reprinted in U.S. Congress, House, Select Committee on Education and Labor, *Hearings on HR 6748, To Provide a Comprehensive Child Development Program,* 92nd Cong. 1st sess. (Washington, D.C.: U.S. Government Printing Office, 1971), pp. 485–537.

14. *Boston Globe,* July 4, 1971.

15. MIT Planning Office, *Op. cit.*

16. Urban Research Corporation, *Industry and Day Care* (Chicago, 1970), p. 3.

17. James March and Herbert Simon, *Organizations* (New York: John Wiley and Sons, 1958); R. M. Cyert and James March, *A Behavioral Theory of the Firm* (Englewood-Cliffs, N.J.: Prentice-Hall, 1963); Joseph Bower, "Descriptive Decision Theory from the 'Administrative Viewpoint,' " in Raymond Bauer and Kenneth Gurgen, ed., *The Study of Policy Formulation* (New York: The Free Press, 1968), p. 103.

18. The literature is rich in models of decision-making in business and, more recently, government. Yet models for university decision-making are not available. Recent upheavals should lend urgency to understanding the university decision-making process. Similarities in substance and nomenclature suggest that a fully developed model based upon the sketch proposal in this section might have widespread applicability. Cf. Joseph Bower, *Managing the Resources Allocation Process: A Study of Corporate Planning and Investment* (Cambridge, Mass.: Harvard Graduate School of Business Administration, 1970), pp. 1, 285–286; Graham Allison, *The Essence of Decision-Making* (Boston: Little, Brown, 1971), p. 13.

19. Tufts University is composed of three undergraduate colleges: the College of Engineering, which is coeducational, the College of Liberal Arts, which is comprised of men, and Jackson College for Women. Presently there is no distinction between the latter two colleges.

20. Louise Gross and Phyllis MacEwan, "Day Care," NUC Women's Caucus, 1970.

21. This office had been created following a confrontation caused by the lack of minority employees on a federally financed construction site. See Antonia Chayes, Christopher Kaufman, and Raymond Wheeler, "The University's Role in Promoting Minority Group Employment in the Construction Industry," *University of Pennsylvania Law Review* 119, no. 1 (November 1970).

22. James Coleman et al. *Equality of Educational Opportunity* (Washington, D.C.: U.S. Government Printing Office, 1966).

23. Joseph Featherstone, "Kentucky Fried Children," *New Republic*, September 12, 1970; Urban Research Corporation, *op. cit.*

24. In 1971 the Tufts Center had fifty-four children, including fifteen spaces alloted to children receiving AFDC. Tuition fees plus those from the welfare contract covered costs.

25. American Academy of Arts and Sciences. *The Assembly on University Goals and Governance* (Cambridge, 1971), p. 8; Frank Newmann, *Report on Higher Education* (Washington, D.C.: Office of Education, Department of Health, Education and Welfare, 1971), ch. 4; Carnegie Commission, *Less Time, More Options* (Berkeley, Calif., 1970).

26. Frances Horowitz and Lucille Padden, "Effectiveness of Environmental Intervention Programs," University of Kansas, 1970 (mimeo).

27. "A Proposal to Study Evaluation Criteria for Child Development Projects and Programs," Part I (Cambridge, Mass.: Abt Associates, 1971).

28. Henry Etzkowitz and Robert Zeflert, "Strategy and Tactics of Institutional Formation," Washington University, St. Louis.

29. Keith McClellan, "Strategy for Day-Care Cost Analysis," paper presented to the Day-Care Policy Studies Workshop, Arlington, Virginia, May 1971. McClellan outlines a useful strategy for *cost* analysis but with his Abt colleagues firmly believes that cost-*benefit* analysis is not yet feasible.

30. Bettye Caldwell, "Day Care: Pariah to Prodigy," *Association of American Colleges for Teacher Education Bulletin*, vol. 14, no. 2.

31. Evelyn Pitcher, "A Day-Care Education Center at Tufts University," Tufts University, Medford, Mass., 1971.

8

CHILD CARE

AND WOMEN'S LIBERATION

ELIZABETH HAGEN

At the Congress to Unite Women, held in New York in November 1969, a national coalition of women's groups called for free twenty-four hour child-care centers, employment equality for women, and the deduction of child-care expenses from taxable income.

Good child care is crucial to women's liberation. It is also crucial to the better education of small children, to the easing of stresses produced within the nuclear family, and to the greater integration of the lives adults lead with the lives their children lead. As the women's movement gains momentum, the responsibility for raising small children is coming back, step by step, to the community at large.

Feminists believe that the rearing of children is a matter for the entire society, not simply for individual parents or, more narrowly, for individual mothers. Rosalyn Baxandall, one of the founders of the Liberation Nursery on East Sixth Street in New York City, says, "Just as education from six up was taken out of the home, education from birth to six should be too. Families may have been able to do it (educate their children) several generations ago, people lived in extended families then, but I don't think families can do it now."[1] The poverty of the individual family setting compared with the richness of a good day-care center in terms of the stimulus, skill, and dedication of the staff has to be admitted, but the controversy over day care extends far beyond mere comparisons of equipment and staffing.

Day care is not to be compared with nursery school. Good day-care centers are trying to provide the kinds of facilities that are already offered by some high quality nursery schools, but the emphasis lies in providing children with an influence and an environment that will be as powerful and as important as their home setting, not simply a supplement to their mother's care. Susan Edmiston writes, "Day-care people see the center or school not as a foreign, outside influence but ideally as an environment created by the parents themselves, acting in community." [2] She also points out,

> Motherhood, as recent generations have known it, is tottering. Traditional child rearing has been strictly tête-á-tête; one mother totally engrossed in and at the service of one child, one child totally involved with one mother. A sacred obligation, constant and jealous, it required the all-but-perpetual attendance of the mother. An occasional respite when babysitter or mother-in-law cared for the child was acceptable but still guilt-inspiring. Going off to work and abdicating to a mother-surrogate was inexcusable. Those women who did so were always subtly, or not so subtly, tainted. The *best* mothers didn't. The assumption was that at any and all times, mother knew best and mother *was* best. Mother and child were locked in an eternal embrace. [3]

In *A Journal of Female Liberation* Lisa Leghorn writes,

> The importance of the private raising of children lies in the services it performs for the male. The family institution ensures that men are not responsible for the drudgery involved in raising kids privately. Women take care of all that. Yet children, mostly male children, are the future of the man, his immortality. That is why they must follow his footsteps, carry on his occupation or concern for him. His sons will carry on the "family" name, his name. And they will inherit the "family" property, mostly his. This ensures that man's fear of death can be dealt with. He never really dies as long as his children are perpetuated in his image. We have seen what this security for the male has done for the women and children. [4]

Most mothers make an effort at some time to socialize their children, either through play groups or nursery schools or in some other way. Many of them make far greater efforts for their children than they do for their own social lives, but such self-abnegation is entirely acceptable in present-day society. Few children whose mothers can afford it miss at least two mornings or afternoons a week at some form of nursery school. Parental responsibility for teaching a child to associate with other children is accepted generally by most people. However, where parents take care to include their children in such groups, they will consciously or subconsciously ensure that the activities do not represent any kind of parental "out."

Nursery schools are not structured to give parents the opportunity to work at part-time jobs or to take part in any kind of regular extradomestic activity. They open after office hours and end two and a half or three hours later, cutting up a morning or afternoon irrevocably for mothers. Parents usually contribute time to cooperative nursery schools on a strictly sexist basis; fathers, around eight hours a year for carpentry and repairs; mothers, two and a half hours or so every three weeks. One Princeton mother says, "I clean three times as many johns as I do at home, and that is my total contribution."

Driving to and from nursery school, even with car pools, takes up considerable blocks of time. Under this system children are dipped briefly into another world and fished swiftly out again. Before their mothers have time to turn around, Johnny and Janie are home once more.

According to Louise Gross and Phyllis MacEwan in a New University Conference Women's Caucus publication,

Historically in the United States, full day care programs, as contrasted to half-day nursery schools, have been provided in periods of economic stress—in, wartime or during the depression—when women were required in the work force. These programs were created primarily to serve the corporations that needed woman-power, not as an educational and social opportunity for children. Although wartime day care centers often became educational opportunities for children, their rapid closing following World War II was a clear indication that these centers had not been organized primarily to benefit children, or even to liberate women. Rather, they had been organized to facilitate the carrying out of needed production.[5]

The crux of the women's liberation view of day care is the concept that mothers need to be liberated from confinement to the home and so do children. Mothers need work and contacts outside the domestic setting and so do children. Mothers need to be released from the stifling tête-á-tête of traditional motherhood and so do children. As mothers need to find stimulating, educational environments in which to function, so do children. Thus day care becomes a dynamic concept, not simply a stopgap to cover the years between breastfeeding and kindergarten, or the hours between the start of work in an office and the beginning of domestic chores.

Early in 1969 a group of women in Washington, D.C., began to meet with the intention of working out a plan for day care. From a preliminary play group the enterprise grew to the status of a school with plans for a full day care all year round. By February 1971 the group had found a ramshackle house, and twenty-five children from the ages of ten months to three years were coming regularly to the school, Children's House. It opened from 9 A.M. to 5 P.M., Monday through Friday. An approximate ratio of one adult to three children was maintained, and children were signed up for any combination of hours. Adults came in morning or afternoon shifts, working one-third of the time their children spent at the school. Men and women worked equal time, and couples with two or more children worked the same amount of time as those with only one child. Single parents worked only one-sixth of the time spent at the school by their children. The intention behind this arrangement was to avoid traditional social discrimination against single parents or, for that matter, parents with two or more children. Many nonparents also came to the school. Signing up for different hours each week was possible. This policy kept the center flexible enough to meet everyone's needs.

Administrative jobs were rotated weekly through the list of adults in the group. These included scheduling, buying supplies, laundry, and cleaning up once a week. A standing finance committee signed checks, paid bills, and kept records straight. A standing equipment committee bought some toys and furniture and made others. In fine weather the group took trips to museums and parks and went for walks. At meal times the children ate finger foods like cheese, apple slices, and meatloaf, sitting down, standing, or wandering around as they wished. In the afternoon many of the children slept.

Provisions for licensing day-care centers in Washington are stringent, and no attempt was made to meet the requirements, which included regulations against signing up children under six months old, boys and girls sleeping in the same rooms, and various safety code regulations. Two toilets instead of the statutory five placed the school totally outside the official acceptability field.

Parents involved in the Children's House tended to be white, college educated, and politically left of center. Theories and a philosophy of education emerged during the first eight months of operation that reflected these particular origins. Parents believed that children should have the freedom to grow, explore, learn, and be, which meant providing a secure, interesting environment. It also meant respecting childrens' ideas, emotions, and business. Little structure was imposed on the children. They were free to do whatever they pleased.

Marcia and Norma wrote in *Off Our Backs*, "Parents should not always be the only important people to a child, even an infant. We must let our children be free to develop very special relationships with other adults and . . . we have the time and openness to develop the same loving concern for other children that we have for our own." [6] Marcia and Norma also mentioned that parents were nearly always shocked the first few times when their leave-taking went unnoticed. They tended to say several goodbyes until their children shed a couple of propitiatory tears, thus releasing the anxious parents.

The way in which this center is governed reflects the struggles of the women's movement to achieve humane methods of operation in an inhumane world. Marcia and Norma described this center's means of governing:

In place of taking votes or establishing rules, the group has tried thorough on-going discussions to reach a consensus that is satisfactory to everyone. Issues are always open for more discussion and new consensus. We have always hoped to relate to each other in an open, honest way that is supportive. This involves the very difficult tasks of really listening to each other, being tolerant of where others are at, and still expressing and working for what you think is correct. All of us have been concerned about solving our problems without making the meetings painful to attend. Sometimes, issues have been settled, but it hasn't always worked; the group is too large. We tend to polarize at meetings. Meetings are repetitive and we often avoid dealing with issues. We don't yet know or trust each other and there hasn't been much true struggle or true support for each other. We find it very difficult to talk about child rearing without getting defensive about our own children and practices. Because we haven't been successful as a group, a woman whose husband wasn't doing his share would have to leave the group or fight him alone rather than having the group deal with him as an irresponsible member.[7]

The Children's House is committed to fighting sex-role stereotypes.

We want children to know men as well as they have traditionally known women. To see that all people have their good and bad, strong and weak times and to see that all jobs, all emotions, all ways of being are as appropriate for one sex as the other. We don't want our children to grow up thinking that little boys become whatever they want, while little girls grow up to be mothers. But, it is hard for people to change and we have not really dealt with much of our own destructive socialization.[8]

Changes in the life styles of the adult members of the Children's House have gradually become necessary. One man fought for a two-thirds time contract at his teaching job in order to work at the center. Two women asked for release time when being interviewed for jobs. Men cooked, washed dishes, and spent much more time with their children. The influence of this philosophy of human liberation is beginning to spread as the rigidities of traditional living and employment patterns are increasingly recognized as destructive and dehumanizing.

In the summer of 1969 a cooperative day-care center was started in Bloomington, Indiana, by members of women's liberation. It used a Unitarian Church Sunday School building and was open from 8 A.M. to 5 P.M., Monday through Friday, with fifteen or twenty children attending all or part of the time. A second center was started in January 1970 on the ground floor of the Unitarian Church's main building. Prospective members visited the centers and attended meetings. Whether they were admitted depended on mutual agreement between themselves and the current participants.

Work has also begun on a cooperative day-care center at Indiana University. The members of the Indiana University Day-Care Campaign felt that the principle of parental control was important since it provided resistance to some of the restrictive practices currently used in many centers organized by money-making organizations or controlled by institutions. Many such preschool centers stress routine, a mass-managed schedule, the learning of obedience to authority, sex-role identification, and individual competition. The Day-Care Campaign at Indiana states: "Parents who are intimately and daily involved in determining the policy and the activities of their own center will not allow a philosophy to exist there which is alien to their own developing experience and understanding of what is best for people." [9]

After more than a year's work the Women's Rights Committee of the Hyde Park-Kenwood, Illinois, Community Conference opened a non-profit, parent-controlled child-care center for working mothers in the Sunday School facilities of the Church of St. Paul and the Redeemer. It

seems appropriate that the irascible saint should at last have an opportunity to make amends to women! The group had to sign a contract according to which it took full responsibility for the center's operation, including insurance, and the church incurred no costs. Changes were
found to be necessary under Chicago's code requirements for child-care
centers. Various requirements for state and city certification had to be
met. The final go-ahead came when a building department official
checked his record book and told the women that except for one lighting requirement, their license had been approved back in April.

The Spokeswoman described the financial arrangement that made this
opening possible. Mary Houghton, an officer of a community bank, prepared sound business projections, a five-year budget, and a profit-and-loss
statement. Her figures showed that the group needed $5,000 to launch the
center and that a $4,000 loan could be obtained if the group raised
$1,000 and solicited guarantees from numerous community figures for
the amount of the loan. A "Bids for Kids" auction was held on October
25, which raised $1,800. A Hyde Park-based business, the Urban Re-
search Corporation, gave an initial guarantee for $1,000. Further loans
were still being processed in December 1970.[10]

The group planned to hire a full-time director at a salary of $10,000
and start with twenty children. Full capacity would be fifty-three children. Teachers would be hired at a ratio of one teacher to ten children.
Cost worked out at sixteen dollars per child per week. No parent would
be charged more than cost. Some half-day and afterschool children
would come to the school, but basically the idea was to provide care
from 7 A.M. to 7 P.M. for the children of working parents.

In September 1970 the Central New Jersey Chapter of the National
Organization for Women (NOW) opened a day-care center after a he-
roic summer of hard work and persistence. It had no funds, no backing,
and no premises until the last moment, when Princeton University came
through with the offer of four large classrooms and a teachers' lounge in
one of its campus buildings. This building already housed a twenty-
year-old cooperative nursery school, and at first there were some hard
feelings between the two establishments, both alike in some ways, but
totally different in long-term objectives and methods of functioning. One
of Princeton's deans of women gave support and interest that turned
out to be crucial for the success of the venture, and the school now runs
with a maximum admission of forty-five children, cared for by a full-
time director, three full-time teachers, and six half-time teachers.

The Princeton University NOW Day Nursery is open from 8 A.M. to
6 P.M., Monday through Friday, and has recently raised its weekly fees
from twenty-eight to thirty dollars for a child attending full-time. Princeton University gave its backing to the extent of free space and utilities
for another year after one school year of operation. All salaries, equipment, taxes, running expenses, and food are paid for out of income. Princeton University recently granted a loan of $2,000 to cover capital

expenditure, and there are four half-scholarships available for university-affiliated children, to be taken up in September, the start of the new school year.

The basis on which the school is run represents a compromise between the idealism of the Children's House and the conventional methods used by commercial or state-funded day care. A board of directors makes all policy decisions and hires the director and teachers. This group consisted at first of members of NOW, some parents, and interested outsiders. Recently there has been considerable pressure from some parents for complete parental control. Members of NOW are concerned that the feminist influence in the school may disappear under pressure from a conservative community, but there can be no control of the nursery, for tax exemption reasons, by any political group, and elections of board members will automatically determine what sorts of influences parents want in their school. It is ironic that public belief in day care will probably come sooner than public understanding of feminism!

At the outset salaries ranged very little, only $2,000, from top to bottom, and teachers seriously debated whether to introduce an egalitarian system of equal pay for all, regardless of previous experience and length of service. In the end it was decided that salaries would have to vary, and recently a new full-time director was hired to replace the part-time one who bore the brunt of the school's opening troubles. The new director's salary will be $10,000, which is a $2,000 raise over the highest salary payable when the school opened. Many teachers felt resentful when this decision was made, and their point of view is certainly understandable.

Certification seemed very important to the founders of the school, and the conditions of the New Jersey state administration included 5,000 dollars worth of additional fire alarm equipment, for which Princeton was willing to pay, and certain qualifications in teaching and early childhood education for the director and class teachers. After a year of negotiation and debate the school is now duly certified, and a source of worry and dissension among some parents has been removed.

The Princeton University NOW Day Nursery operates within the framework of society as far as possible. Salaries are at least comparable with those of the public-school system; certification has been obtained; the structure of the school is hierarchical in appearance; and administratively some attempt is made to conform to business patterns. Nevertheless, there are many ways in which the school community tries to pursue innovative paths within this structure. Consensus is very important. Board meetings are usually held weekly and last long into the night, with emotions close to the surface and a minimum of arbitrary decision-making. Teachers try to talk to parents at length about their teaching philosophy (as yet unstated in any official form) and try also to alleviate parental anxieties as far as they can.

Not all parents use the school in order to work at full-time jobs; some use it as a form of nursery school and suffer guilt at their deviation from the social pattern of optimum motherhood. The difficulties in group functioning experienced in the Children's House are mirrored in the Princeton University NOW Day Nursery. It has become increasingly necessary to develop a statement of educational policy in order to make it clear to parents and teachers at the beginning of their association with the school that some beliefs and practices shared by the founders of the school are unlikely to change even under considerable fire from dissenters. Plenty of parents feel insecure at the youth of the teachers, at their informality and lack of "talking down" to children, at their willingness to allow conflict and stress in children's day-to-day encounters rather than to intervene immediately in any kind of difficulty, in the name of protection.

Thumps and bashes inevitably result when children inadvertently or "vertently," as one teacher put it, rub up against one another, and teachers vary widely in their attitudes to this problem. Some feel that children, like adults, have to learn to handle their differences with other people and that teacher intervention ought not to be instant or too obtrusive when trouble occurs. Others feel more protective. Elements of guilt and insecurity on the part of parents have made this a sore subject from time to time.

Playground surveillance has also been controversial. What one parent perceives as normal adventurousness, another might see as wild risk-taking. One board member expressed her feelings thus, "When my child was small she went to a dear little group in Syracuse run by a motherly old Granny who made quite sure that all the children were absolutely safe in every way." From the motherly old Granny in Syracuse to youthful feminists in blue jeans is a far cry and a major shift in philosophical approach.

Where possible, older and younger children are mingled in the same classroom. Older children frequently enjoy playing with "baby" toys and returning to their own babyhood. They are intrigued with infant toilet training and feeding, fascinated with the whole process of growing up in another smaller child. Often they can be tremendously protective and helpful. Child-to-child teaching is an important feature in the day nursery program.

The day-care concept requires that children be allowed to pace themselves through a day. There are no play-group scheduled activities chasing each other through a three-hour period with ruthless regularity. The day nursery is home, and it is important that home should be exciting, but not regimented. Recently discussed was a suggestion that a small storeroom should be used as a privacy room, where children could occasionally get away from one another. Some parents immediately raised the objection that privacy means a teacher might not have her eye on the private child at a crucial moment, and "anything might happen." Such insecurities are very much a part of parental feeling and have to be given due weight.

Skills are taught implicitly at the learner's pace. Teaching is child-oriented rather than teacher-oriented. Teachers say, "We plant six beans. That's Math and Science together!" Learning is integrated, and one experience arises out of another in a natural progression. Children who show special aptitudes are encouraged to progress and explore further, but no judgments are made about slow learners or reluctant mixers.

Sex stereotyping is consciously avoided in the Princeton University NOW Day Nursery. Johnny does not have to learn to be "manly," to hold back his emotions and take care of Janie. She, on the other hand, does not have to be "feminine," weeping and helpless. She learns, as Johnny does, to be a human being. They both get dirty, make noises, pummel each other on the floor, play with dolls, fabrics, clay, paints, tools, and blocks. Teachers at one meeting complained that *Mother Goose* was the most sexist book on the school's shelves, but also the most popular. It is hard for a feminist to tolerate, "Peter, Peter, pumpkin eater, had a wife and couldn't keep her, he put her in a pumpkin shell, and there he kept her very well!" Nevertheless, the demand for *Mother Goose* has thus far overridden staff objections and a cleaned-up version gets read as often as ever.

In this school as in other feminist day-care centers, children are not specifically prepared for entry into the public-school system, on the reasoning that this system still conflicts with modern educational theory in many ways and violently conflicts with feminist ideology. Thus teachers hope that during a child's few years of day care she will develop an independent mind and intelligent judgment, which will enable her to cope with the exigencies of the public school or any other environment. In this respect only, perhaps, women's liberation shares a principle with Saint Ignatius Loyola, who offered, by teaching a child for the first seven years of his life, to render him impervious to corruption from then onward.

There are ways in which the school suffers from its determination to operate within the letter of the law. Expeditions are not made outside the school grounds because insurance liabilities would be too great in the event of an accident. A certain amount of flexibility in the kinds of activities that can take place is lost. Teacher substitutes are hired from all walks of life, however, and judged useful or not on their ability to relate to children in the classroom situation. Men are frequently hired as substitutes, and one has taken a position as a teacher during the summer program. Another stereotype is on its way out of the minds of children, teachers, and parents.

Criticism has been aimed at the school for its all-white, middle-class character. This is hard to avoid, given the level of its fees and the nature of the surrounding community. Good day care is too expensive for more than a small percentage of the population to afford. At the same time the teachers' wages are too low to compete with the public schools. This summer the school lost two of its best full-time teachers to the public schools, where the pay is better and there are summer vacations. The nearest approach to a summer vacation the school is able to afford is one "mental health day" in every month that does not have a public holiday. This is in addition to normal vacation time (one week).

The question of racial integration has been much discussed, and there is a possibility of amalgamating with one of the two other day-care centers in Princeton in order to absorb children whose day care is funded by the Office of Equal Opportunity. The overall size of the school cannot be increased without raising costs per child, so the absorption of the OEO children would have to take place gradually according to the waiting list, without displacing existing applicants.

The philosophy of the school is an evolving concept whose shape and quality has become increasingly well defined since September 1970. It arose almost spontaneously from the group of young teachers who drew their ideas from a common habit of communication and tolerance. This habit has been fostered in the children, who are constantly invited to verbalize their ideas and difficulties, to express their emotions, and to allow others to do the same. It is a revelation to hear one four year old say to another over a disputed toy, "Let's negotiate!"

This philosophy is vulnerable to attack from members of the community who have been raised in hard schools and who regard the willingness of teachers and board members to hear all points of view as a weakness. Teachers who genuinely attempt to give parents as well as children their attention at the end of every long day find themselves exhausted mentally and physically. There is a need for protection for teachers, and one of the functions of the new full-time director will be to screen off a little of the pressure. This requires a delicate balance between giving parents the reassurance they badly need and overloading teachers to the point where they lose their classroom effectiveness.

Day care is still a highly controversial issue in which many questions have not yet been answered. Little is known about infant day care, except that it is far too expensive to operate on a fee-paying basis, as the ratio of adults to children must not be less than one to four, preferably one to three. The Princeton University NOW Day Nursery had to drop its infant class for financial reasons. Nevertheless, the need for infant day care is undisputed.

All day-care centers have an educational function to perform for the community at large. Conventional preopening "need" surveys are useless. People will not commit themselves to day care until they have an existing day-care center to study and observe over a period of time, and they will not base career plans on a day-care center unless there has been an opportunity for them to use and test it for quality and relevance to their needs.

Above all, a change in social attitudes has to take place on a large scale. This is the battle we are all fighting with the women's movement. The right of a mother to work outside the home has to be as accepted as the right of a father to do the same thing. For this reason it can be argued that cooperative day care is a kind of copout. Many parents work long hours or commute and are not free to work in day care. They have to find places where their children will receive good educative care from the best of teachers without any parental contribution other than a financial one. In the West Village Cooperative Day-Care Center in New York, which serves a large number of children under two, mothers who work full time pay twenty dollars a week; others work in the center one full day a week. Both give four dollars a week for supplies. The center has about thirty children, with four or five under a year old, and stays open from 8:30 A.M. to 6:00 P.M. This center, as well as the Discovery Room on the upper fringes of Harlem, which has thirty-two children and owes much of its educational philosophy to the British infant schools, is funded by the City of New York, after an initial start with very little money and a great deal of energy and enterprise.[11]

According to Susan Edmiston, the cost of good day care is generally estimated at forty dollars per week. Some industries, as discussed by Susan Stein in Chapter 14, have set up day care for their employees. Women's liberationists generally agree that there are built-in drawbacks to company day care. These include increased dependence on the corporate structure for individual families, mothers being tied to the firm at which their children attend day care, and the reluctance of most firms to grant the use of day-care facilities to male employees on the assumption that it will invariably be the female employee who takes the child to work with her.

Rosalyn Baxandall has further noted,

Too often it (industrial day care) ties women to lousy jobs. The industries that are setting them up are the ones that can't keep employees, they're bribing the women. Day care centers should be in the neighborhoods where people live. It's not a good thing to drag a child to work with you on a subway. Besides, parents ultimately don't have too much to say about day care when a big corporation is controlling it". [12]

The Day Care and Child Development Council of America has launched a program to help groups to develop early childhood services in their communities. The objectives are to establish universal publicly supported, community-controlled day-care systems. It seems clear that only under such a system will responsibility for young children be laid where it belongs, on the shoulders of the community at large. Privately financed day care is prohibitively costly and subject to continual harassment arising from the attempt to reconcile state regulations (designed to control welfare day-care institutions) with makeshift housing arrangements necessitated by lack of funds. Day-care centers ought not to be run in church rooms, tumbledown houses, basements, and campaign offices. They require planning and construction geared to the purpose for which they are to be used, specifically, as very special environments created for very special people. If schools for children over five are important enough to be planned and designed, so are schools for children under five.

Louise Gross and Phyllis MacEwan expressed the problem well:

> We feel the differences among the existing day care centers reflect a conflict in values and attitudes towards human development. This conflict in the care and education of young children is directly related to conflicting values and attitudes expressed in the economic and political behavior of adults. Values in competitive enterprise and individual rather than social achievement, respect for private property, adoration of the nuclear family, are attitudes nurtured in childhood and expressed in the adult society. We must be clear about *our* goals for children and recognize that they are in conflict with those institutions, corporations, and universities from whom we are presently demanding day care services. This implies that when we make demands for day care they should be solely in terms of space and money. The corporations and universities should have no control. In defining our position, we need to become aware of how educational goals become implemented within the day care center. We need to be aware of the meaning of curriculum, the teacher's philosophy and style, the role of parents and other related issues.[13]

The primary need in the child-care field is for greater clarification of what constitutes a good day-care program. On this foundation structures for financing and developing day-care centers can be built. The childcare concept goes hand in hand, as we have seen, with the concepts of female liberation, male parental responsibility, and community responsibility for small children. It also goes hand in hand with a new view of the coming generation and the lives this generation will lead.

Let us hope that the small community centers described in this chapter represent the grass-roots beginnings of a great American movement.

NOTES

1. Rosalyn Baxandall, quoted by Susan Edmiston, "The Psychology of Day Care," *New York Magazine*, April 6, 1971, p. 39; cf. Rosalyn Baxandall, "Perspectives on Our Movement: Cooperative Nurseries," *Women: A Journal of Liberation* (3011 Guilford Ave., Baltimore, Md. 21218) 1, no. 3: 44–46.

2. Edmiston, *op. cit.*

3. *Ibid.*

4. Lisa Leghorn, "Child Care for the Child," *A Journal of Female Liberation* (371 Somerville Avenue, Somerville, Mass., 02143), no. 4 (April 1970): 29.

5. Louise Gross and Phyllis MacEwan, "Day Care," Women's Caucus, New University Conference, 1970 (mimeo).

6. Marcia and Norma, "The Children's House," *Off Our Backs* (P.O. Box 4859, Cleveland Park Station, Washington, D.C. 20008) 1, no. 17: 9.

7. *Ibid.*

8. *Ibid;* cf. Lenore J. Weitzman, Deborah Eifler, Elizabeth Hokada, and Catherine Ross, "Sex-Role Socialization in Picture Books for Preschool Children," *American Journal of Sociology,* 77, no. 6 (May 1972): 1125–1150.

9. Day-Care Campaign, Jeff Sharlet Chapter of New University Conference with the Aid of the Indiana University Student Government (contact Jean Romsted at Student Government Office 7-94-15, Indiana University, Bloomington, Indiana); see *The Spokeswoman* (an independent women's newsletter published at 5464 South Shore Drive, Chicago, Ill. 60615) 1, no. 9 (February 1971).

10. *The Spokeswoman* 1, no. 7 (December 1970).

11. Edmiston, *op. cit.,* pp. 39–40.

12. Rosalyn Baxandall, quoted in Edmiston, *op. cit.*

13. Gross and MacEwan, *op. cit.*

PART III

Cross-National
Perspectives on
Child Care:
A Look Abroad

9

WHAT OTHER

NATIONS ARE DOING

PAMELA ROBY

Authors of Part I of this book have analyzed basic changes in American life that have created a critical need for additional child-development services throughout our nation. In the hope that groups creating our needed child-development programs will build upon existing knowledge, authors of Part III described varied past and present U.S. child-care programs and analyzed their experiences with these programs. In addition to telling us much about "what works" in our existing, numerically limited, preschool programs, their chapters highlight numerous questions and issues underlying most discussions concerning the construction of future child-development programs and policies in the United States.

Other nations have already confronted or are now confronting many of these early childhood issues. Therefore, in this part educators, sociologists, and other early childhood development specialists describe and critically examine current and projected child-care policies in Sweden, Finland, Hungary, Norway, England, the Soviet Union, Japan, and Israel. The degree to which the experiences of other nations may be applied in the construction of U.S. policies is partially dependent upon their economic and social similarities to the United States. Therefore, only highly industrial nations with high or rising female labor force participation rates are represented in this volume. For this reason also the authors describe the purposes and philosophy of their nations' early childhood policies before examining the programs and policies themselves.

Although the United States is the world's wealthiest nation, other nations have forged far beyond us in the provision of early childhood services. While only one-tenth of our states offer kindergarten to all their children, Israel, whose 1969 gross national product per capita was only $1,663 as compared with the U.S. GNP of $4,664,[1] provides kindergarten for all its five years olds, child care for 50 percent of its three and four year olds, and comprehensive health care through neighborhood Mother and Child Clinics for 90 percent of its infants and their mothers. The Swedish government also now provides a wide range of services to all its youngsters and plans to provide child-development centers for 85 percent of its preschoolers by the mid-1970s.[2] In addition to extensive social and health services, Hungary supplies nurseries for half of its three to six year olds and creches for others. Equally significant for Americans concerned with infants and preschoolers are other nations' program and policy innovations.

Americans entering Scandinavian or Israeli children's centers are immediately struck by their hominess.[3] Curtains hang at the windows, and framed pictures are on the walls. These "day homes" or kindergartens generally have three or four rooms that are one-third to two-thirds the size of the typical large single American kindergarten room. Children who wish to get away temporarily from the bustling, noisy play activity of one room may go to another room with a carpet, a low couch, pillows, and stuffed animals. There they may look at picture books, work with puzzles, or simply enjoy quiet thoughts and privacy.[4] Throughout Israel and in many Scandinavian areas most children's centers are housed in separate buildings or "children's houses." In other Scandinavian areas the children's centers are housed on the first floor of apartment buildings. In either case one room usually opens out onto an enclosed play-yard and garden. Children go in and out at will. Outdoors they may care for pets, swing, garden, build with blocks, or play in the sandbox. Inside each child has one or two drawers of his or her own in which to keep paintings and other art work they have done as well as personal toys and other items. When nap time arrives teaching assistants bring out cots. A sense of stability is built into the one and a half-hour nap times by each child's having his or her cot set up in the same spot day after day.

Descriptions of the children's activities and menus for their meals are posted in the foyer of the "day homes" so that parents may know on a day-to-day basis what their children are doing and eating. Parents are also encouraged to discuss the kindergartens' activities with their children and to discuss problems their children may be having with the teachers. Because of the difficulties involved in parents bringing their children with them to work, the children's centers of most nations included in this part are now primarily located in the neighborhood of the children's homes rather than near their parents' work places.

American visitors to these nations are also struck by the seriousness with which their hosts explain that child-care center policies should be considered only in the context of a *comprehensive* social policy for the promotion of children's well-being and development and the well-being of their parents. Many nations' provision of comprehensive children's services stands in marked contrast to the American practice of attempting to assist low-income children by providing one isolated service after another rather than a comprehensive program of services: one year we offer a social-service program; the next a dose of Head Start and other compensatory education programs; then, while easing up on education, we emphasize health programs. Mounting vast evaluation research projects with each of these new programs, we too often conclude that each new effort is a failure because the children involved in the program—who need not one isolated program or service but an environment that includes stable, adequate family income; decent housing; nutrition; health care (including prenatal care for their mothers before their birth); education; supportive parental services; and love—do not do remarkably better in grade school than do other youngsters who have not been given the single experimental service. The contrasting comprehensive children's services of Sweden, Israel, Hungary, and other nations are built upon the belief that all human beings including children have the right to a decent level of well-being as well as the belief that the nation must concern itself with the welfare of its children today because they will be its citizens tomorrow.

Because many nations hold that child care should be provided only within the framework of *comprehensive* child-development services, this part will not only contain careful descriptions and examinations of the operation, administration, financing, and goals of other nations' infant-care centers,[5] early childhood development centers and family day-care services, but will also describe each of the nation's publicly provided children's or family allowances, maternity and paternity leaves, birth or layette grants, maternal prenatal health care plans, infant and early childhood medical care systems, subsidies of nutritious food and clothing for children, and housing grants to families. In addition to describing these provisions, which many nations consider basic children's rights, authors of the chapters in this part will introduce the reader to Finland's and Norway's "park mothers," who provide year-round, daily, supervised playground and park activities, and Scandinavian "child visitors," who look after sick children in the children's homes when their parents are at work.

Other nations, while forging far beyond the United States in providing many children's services, also share many of the child-care problems that the U.S. is currently confronting. All except perhaps the Soviet Union have an urgent need for additional child-development centers. Some nations have taken steps, however, to make planning of child-development centers an integral element of community and national planning in order to make sufficient child-care services available to parents as rapidly as possible. Also most nations are still striving to raise the salary levels of preschool workers and to bring more men (or some men) into the field. All nations have a large fraction of preschoolers without fathers, and all acknowledge that having men on child-care staffs is very important for these youngsters. Sweden is currently attempting to bring men into preschool education by exempting those who work with small children from the draft and by granting them special fellowships, but most planners in Sweden and other nations feel that the wages of pre-school staff must be made more attractive for men to remain in the field. Finally in all nations many persons are working to create sufficient child-care services because they view these services as prerequisites for the establishment of equality between the sexes or for the better utilization of women in the labor force. In those few places where sufficient child care is available, persons are finding, however, that adequate child care brings a society only part way toward the realization of these ends. Not only the provision of child care and the end to educational and occupational sex discrimination, but also males' full sharing in familial and societal childrearing are required if men and women are to be emancipated from their limited sex roles and if women are to be fully utilized in the labor force.[6]

NOTES

1. United Nations, *Statistical Yearbook, 1970* (New York: United Nations, 1971, pp. 603–605.

2. Interview with Mr. Mats Hellström, Member of Parliament and Chairman of the Commission on Child-Care Facilities, Sweden, August 27, 1970.

3. I am indebted to the many teachers and directors in Israel, Denmark, Sweden, Finland, and Norway who cordially showed me their kindergartens and infant and child-care centers.

4. In one kibbutz where I stayed, all kindergartens have four, five, and six year olds. During one hour in the morning and one hour in the afternoon, first-grade lessons are given by one of the kindergarten teachers to those children who are six years old. This system allows children to avoid having to both begin formal learning and cease playing in the kindergarten surroundings at the same time. It also allows four and five year olds to see the six-year-old lesson room and to work in it freely during hours when lessons are not being given. Insurance regulations and laws of many U.S. states require that all children in a class be in sight of their teacher at all times. Such laws preclude privacy and force an entire class to go to the playground or lavatory together.

5. Insurance regulations and state laws currently prevent the establishment of infant-care centers in most U.S. communities.

6. Readers interested in child care and rearing in China are referred to Ruth and Victor Sidel, "The Human Services in China," *Social Policy*, Vol. 2, No. 6, March/April 1972, pp. 25–34 and Ruth Sidel, *Women and Child Care in China: A First Hand Report* (New York: Hill and Wang, 1972); for a report on child care in Cuba, see Marvin Leiner and Robert Ubell, "Day Care in Cuba: Children Are the Revolution," *Saturday Review* 55, no. 14 (April 1, 1972): 54–58. For an examination of child care in Belgium, Canada, France, Germany (FR), Italy, the Netherlands, Sweden, and the United Kingdom, see Organisation for Economic Cooperation and Development, *Working Document: Early Childhood Education*, Center for Educational Research and Innovation, Organisation for Economic Cooperation and Development, 2 rue André Pascal, Paris-XVI°, April 1972, 274 pp.; for England, see Tessa Blackstone, *A Fair Start: The Provision of Preschool Education* (London: Allen Lane, The Penguin Press, 1971).

10

SOCIAL RIGHTS IN SWEDEN
BEFORE SCHOOL STARTS

MARIANNE KÄRRE,
ANNA-GRETA LEIJON, ASTRID WESTER,
ÅKE FORS, OLAF PALME,
MAJ-BRITT SANDLUND, SIV THORSELL

The Beginnings

As early as the 1930s and 1940s, Alva Myrdal, the Swedish politician now famous for her work at the United Nations, called for greater investment in preschool instruction for the good of the children and society. She founded the first training college for nursery school teachers in Stockholm to provide qualified training for the staff of the nursery schools that for pedagogical reasons had been set up by a municipal housing association.

But during the 1950s nothing happened; the authorities responsible simply "forgot" the many modern and perspicacious recommendations that had been made by a government commission in 1946. Only in the 1960s were the ideas from that earlier period given a new lease on life. The child-supervision resources of Swedish society are now being improved quantitatively, primarily to enable more women to go out to work. But the authorities are also becoming increasingly aware of the necessity of such action for the children's sake as well.

The nursery schools, which developed so early in other countries, have not existed for nearly the same length of time or in anything like the same numbers in Sweden, even though our children do not start school until they are seven. For a long time the authorities responsible for the sweeping reforms that were carried out within the Swedish educational system were oblivious to the importance of the preschool years. The time has now come for preschool reform, and a government commission was recently set up for this purpose.[1]

Social Policies for Preschool Children Today

Out of Sweden's 7,843,000 inhabitants on January 1, 1967, 1,752,000 or about 22 percent were children under sixteen years of age. According to the 1960 census, half the families in Sweden with children under sixteen living at home had only one child. Only 16 percent of families with children had three or more children. However, the children included in this last group amounted to a third of all the children in the country. This means that 7 percent of the population between the ages of eighteen and sixty-six were responsible for rearing a third of the next generation. For a long time Sweden has provided extensive protection for the safeguarding of the right of children to good care and upbringing. Each of the 900 municipalities in the country has a child welfare committee charged with the task of working for improved child and youth welfare.

CURRENT FAMILY BENEFITS

Maternity Benefits

Support to expectant mothers, both married and unmarried, is given by protective legislation and by various insurance benefits. A woman cannot be dismissed from a steady job because she becomes pregnant or has given birth to a child. She has the right to six months leave of absence in connection with the birth of a child. She receives free advice and care at the maternity center both before and after delivery. Ninety percent of expectant mothers make use of these centers, and almost 100 percent of the children are given checkups at these centers during the first year. Expectant and recent mothers are also reimbursed for three-quarters of the actual costs of dental care in accordance with an established list of charges. Obstetric care is free. Practically all Swedish children are born in hospitals.

All mothers receive a maternity allowance through the national insurance system. The sum is SK$_r$ 1,080 ($209). In the event of twins the mother receives an additional SK$_r$ 540 ($104). In order that working women can afford to be absent from their work for childbirth, they receive a supplementary sickness benefit from the national insurance system calculated according to income and paid for as long as the woman entirely refrains from gainful employment, to a maximum of six months. A woman with an annual income of SK$_r$ 16,000–18,000 receives SK$_r$ 22 ($4.25) per day.[2]

Children's Allowances

The most important item of economic support to Swedish families with children has been the child allowance. When it was first introduced in 1948, this regular grant replaced an earlier system of tax allowances for children. At present the child allowance is a quarterly grant amounting to SK$_r$ 900 ($174) a year per child and ceases on the child's

sixteenth birthday. Everyone, irrespective of economic status, is entitled
to it.[3]

If a child under sixteen loses one or both parents, he receives financial
assistance from the state in the form of a children's pension allowance from
the national basic pension fund. Adopted children enjoy the same rights
as other children in this respect. The size of the pension allowance is
regulated by the base sum in the national basic pension, which is set by
the government each month at a "standard sum" in accordance with cost
of living. If one parent has died the children's pension allowance
amounts to 25 percent of this base sum. If both parents have died the
surviving child receives 35 percent of the base sum. The pension allowance is paid from the month in which the parents have died up to the
month in which the child reaches the age of sixteen.[4]

Nursing Grants

For children who "by reason of disease, mental retardation, physical
disablement or other defect are in need of special supervision and care
for a considerable time and on a considerable scale," a special disability
grant known as a nursing grant can be paid. The nursing grant amounts
to 60 percent of the base sum for the national basic pension and is paid
until the child reaches the age of sixteen.[5]

Housing Subsidies

Families with low incomes and young children can obtain a housing
allowance, which is subject to a means test. This allowance is paid to
about 40 percent of all families with children, and it is estimated that
about 90 percent of single parents or supporters of children receive it.
Over and above the state housing allowance, a local authority housing
allowance can be paid to families with children if their housing costs
are particularly high. Unmarried mothers or fathers with the custody of
a small child can obtain homemaking loans to facilitate making a home
for themselves and the child.

New mothers who have difficulty in obtaining housing—primarily unmarried mothers—can live for a while at a home for mothers. If the father does not pay the maintenance, then advances are paid to a total of
SK. 1,740 ($329) per year. A child welfare officer is always appointed to
assist unsupported mothers. Single persons with the care of children also
enjoy certain tax relief. However, no special allowances for families with
children are made in taxation.[6]

Public Health Care

Children under sixteen enjoy, through the parents' compulsory insurance, all the benefits of health insurance with respect to hospital care,
visits to doctors, and medicine. Child health services are provided free
of charge through a number of different local bodies. Two types of
child-welfare centers exist in major urban areas: in conjunction with
children's hospitals and under a physician specializing in children's diseases. Child-welfare stations, under the supervision of the county or

municipal medical officer, exist in county districts. These centers offer free medical checkups to children from birth up to school age and at the same time advise mothers on care and upbringing; nurses also visit the homes of families in need of advice and help answer questions. These services also include free inoculation against smallpox, diphtheria, tetanus, whooping cough, and polio.

The basic concern in child health today is to prevent, and treat in time, diseases and injuries that can lead to disablement. The prophylactic activities of the child-welfare centers have meant a great deal to the health of Swedish children and have prevented many handicaps. In 1966 health control was exercised over 99.7 percent of newborn infants, 96.2 percent of one year olds, 59 percent of four year olds, and 55.6 percent of two to seven year olds. The fact that no medical control is exercised over half of all two to seven year olds does not imply that they are never examined by a doctor. At day nurseries and play schools with their own doctors, children undergo regular checkups, while those cared for at home are examined by private doctors in cases of sickness, when the parents are anxious about the child's health.

A public voluntary health checkup of all four year olds in Sweden, with the help of the child welfare organizations, has been approved by Parliament as a possible way of tracing, before school age, the children who may be in need of special care, so that the disruption to their development can be repaired and the child rehabilitated.[7]

Domestic Assistance

Further medical assistance at the child centers is provided by child visitors, who also look after sick children in the home when parents are at work. Child visitor services are provided through municipal "domestic aid committees," but this form of service is not yet offered on any major scale.[8]

CHILD-SUPERVISION INSTITUTIONS

Good facilities for child supervision are of direct significance to the promotion of equality between men and women in the labor market. A survey carried out in 1967 showed that there were more than 300,000 children under ten years of age in Sweden whose mothers went out to work for more than fifteen hours a week. Half of them, about 150,000, were under seven. The same survey revealed that more than 200,000 mothers with children under ten would like to go out to work if they could arrange for the supervision of their 350,000 children. Thus there is a very great demand for supervision facilities for children of employed parents.

Popular opinion used to regard child-care centers as a sort of poor relief for families where mothers were forced by economic necessity to go out to work. Today they are regarded as a service providing the most reliable supervision possible, under the direction of a qualified staff, and teaching children the rudiments of social behavior before school age

(age seven). The popularity of the day nurseries among young parents is shown by the fact that 200,000 mothers of small children desire this form of care for their children.

Communal child supervision is divided into institutions such as day centers on the one hand and family day care on the other. There are also various private forms of child supervision, some of which are subsidized by the state.

There are three kinds of preschool institutions. *Day nurseries* are for the supervision of children aged six months to seven years while their parents are away from home at work or studying. They are generally open from 6:30 A.M. to 7 P.M. on weekdays and until 2:30 P.M. on Saturdays. They are closed on Sundays.

Nursery schools are for children aged three to six for short periods of up to three hours. Common day nurseries are for children with varying supervision requirements and are basically a combination of day nurseries and play school. They are supposed to be able to admit children for periods of varying duration. So far only a few such institutions have been opened on an experimental basis, but the results of putting together children who require both longer and shorter periods of supervision have been most encouraging.

A supplement to the child-supervision institutions exists in the form of *municipal family day nurseries,* private homes that are inspected and approved by the local authorities. Starting in 1969, day mothers, as they are called, received regular pay and social benefits such as holiday pay, health insurance, and pensions. Their rate of pay is by far the lowest in the country. Day mothers can—but need not—undergo a sixty-hour course in child care sponsored by the local authorities.

There are also uninspected, private family day nurseries that parents arrange for by advertising. The great majority of Swedish children who are not looked after by relatives or neighbors are left in homes of this kind while their parents are away at work.° An arrangement of this kind can work very well. But it came as a shock two years ago when a commission of inquiry revealed how much insecurity this could entail for the children: several thousand preschool children had changed day homes several times within a year. The problem was particularly severe with the very smallest children, those whose need of security and stability is greatest. Day mothers were "fed up" after a while or found the work less profitable and more laborious than they had expected.

Parent questionnaires show that the majority of parents prefer institutional care to family day homes, since they can then be sure that their children are cared for by qualified persons, that a check is kept on their health, and that they are provided with correct food, suitable play ma-

---

° Of the 153,386 children under seven with working parents, 9 percent are cared for in day nurseries, 2 percent in nursery schools, 31 percent in private family day nurseries, 5 percent in local authority family day nurseries, 52 percent are supervised in the home, and 2 percent manage by themselves.

terials, and fairly generous space in which to play under the supervision of permanent staff.

Supervision at home by domestic help is becoming increasingly rare. There is very little domestic labor left in Sweden, though there are about 20,000 qualified children's nurses or *stagiaires,* young girls who have to put in a certain amount of practical child care in a family before entering either a college for children's nurses or a two-year infant teachers' seminary. This practical period has been strongly criticized on the grounds that the girls are underpaid and that they can hardly be expected to learn anything merely by being left to their own devices to look after small children while the parents are away.

It was found in 1966 that some 3,000 children under seven years old were left unsupervised while their parents were at work. No further inquiry was made as to how preschool children managed to cope on their own, but in 1968 a widely publicized television program showed how five and six year olds in Gothenburg, the second largest city in Sweden, passed away the time waiting for their parents to come home by shoplifting in the big department stores. After this program the subject of child supervision began to arouse a more vehement interest, and the demand for swift action became more insistent.

The rate of expansion of child-care facilities has in fact risen. In 1963 the government decided to stimulate it by means of a radical increase in subsidies. It is hoped that within a few years there will be places for 160,000 preschool children, 85,000 in nursery schools and 75,000 in day nurseries.[9]

Physical Requirements of Child Centers

Child centers, with the opportunities they offer for activities and contacts with the outside world, have an important function to fulfill. The requirements thus far made of the physical plant have related to such elementary and, of course, very important aspects as spaciousness, rooms at ground level, satisfactory daylight lighting, adequate sanitation, and sensible planning. General regulations of this type have been published by the National Board of Health and Welfare, the ultimate supervisory authority in this field. The Board has also specified in detail how premises should be arranged. It is stipulated, for instance, that there should be an indoor play surface of at least thirty-two square feet per child—preferably more. The outside surface available should be about 110 to 160 square feet per child, preferably more. The minimum total play surface of the nursery schools must be 110 square feet per child.

The Board recommends that child centers should be housed in separate buildings, particularly day nurseries. The Board has published sample drawings for the use of local authorities and others; it has also approved a number of "type solutions" submitted by the manufacturers of prefabricated buildings.

The Board emphasizes that child centers should be as flexible as possible, so that they can be used for other purposes if necessary. Requirements in residential areas shift very rapidly; in new areas there is a great need of facilities for preschool children, but in a few years the emphasis can be on facilities for school children. It must then be easy to adapt centers so that they can be used for older children.[10]

Goals and Activities

The different institutions all pursue the same objective, namely, the "promotion of the children's personal development and social adjustments." They aim to give the children a sense of community, experience of contact with others; and the ability to cooperate with others, take care of themselves, and to a certain degree adapt to the requirements of school life. In theory at least, pedagogics and child care are regarded as identical in Sweden, so that the same standard of training is required for all staff, whether they are concerned with younger or older children.[11] Children are supposed to make their own discoveries with as little direct guidance as possible from the teacher. Thus activities are dominated by free play and free creative work, painting, woodwork, and role playing. Examples of more controlled activities are the "meeting times" arranged at most of the ordinary preschools; the nature of these can differ widely, depending on the teacher, the children, and the interplay between them. For the older preschool children there is a system of "centers of interest," a form of organized group work in a particular field. This type of activity is an attempt to introduce an element of learning and at the same time to train children in more goal-directed group cooperation.

Otherwise, Swedish preschools put relatively little emphasis on structured learning; other than spontaneous learning, the primary concern is to promote more generally the children's emotional, social, intellectual, and physical development.[12]

Staffing

Day nurseries are staffed by nursery school teachers, children's nurses, and in some cases instructors in child care. The number of children per staff member varies with different age groups. The principle is that the younger the children, the fewer children there should be per department (age division). In day nurseries two nursery school teachers should be attached to each department, in addition to the principal. If nursery school teachers are unobtainable, qualified children's nurses can be employed. Apart from kitchen staff and similar help, there should be one staff member to every five children. If the day nursery has a department for babies (six to twenty-four months)—which is where the nurses normally work—then the personnel requirement is one staff member to every four children. Further assistance from the child centers is provided by *child visitors*, mentioned above, who look after sick children in the home when their parents are at work.

Nursery school teachers study for two years at state nursery school teachers' training colleges. The training of children's nurses (nursery nurses) can comprise either a one-term course in the care of small children, plus practical experience among younger preschool children, or a thirty-four week course covering the care of both babies and small children. Instructors in child care, who are qualified to become the principals of homes for babies or of day nurseries with special departments for babies, undergo special training for three years.[13]

Administration and Finances

Preschools (both day nurseries and nursery schools) are under the supervision of the National Board of Health and Welfare, which is under the jurisdiction of the Ministry for Health and Welfare. The regional authority is the county administration and the local authority is the municipality.

A number of large municipalities, which run preschools on a large scale, employ consultants or inspectors who are responsible for coordinating the activities of day nurseries, nursery schools, free-time centers, and childminding, including the training and administration of the assisting staff. In Stockholm the mental health organization also has a preschool team to assist staff in the field of mental health.

A preschool may be under the jurisdiction of the local authority, an association, a company, or a private association or person. Most day nurseries are under the jurisdiction of the local authority. Nursery schools too are usually under the local authority, but quite a few are run by associations, in most cases with the help of local authority grants. Anyone setting up a child center can obtain a state "starting grant" and cover most other initial costs by a state loan.

Starting grants are available for premises arranged so that they can be used for the supervision of children throughout the day, or for at least five hours a day. The creation of day nurseries is thus subsidized by the state, and the same is true of institutions functioning as both day nurseries and nursery schools. Generally speaking, child centers receive both a grant and a loan if the layout and fixtures of the building are planned for group activities by children, and all the children accepted can stay there for a minimum of five hours a day. This means that the institution must offer hot meals and facilities for rest and sleep. If these conditions are met, nursery school departments are also eligible for grants and loans.

Child-center premises should be planned in consultation with the Board of Health and Welfare; the Board also determines the number of places, which must always have a given relationship to the space available and its layout. The State Inheritance Fund also issues grants toward the establishment of nursery schools. If the nursery school is in a residential area eligible for state loans, then a state housing loan can be obtained.

In the case of state grants toward running expenses the requirement again is that activities cover at least five hours per child per day. It is also assumed that the children will be under the supervision of qualified staff and that the premises will be suitably equipped. If a given institution, for instance, has both day nursery and nursery school departments, then a grant for current operations will be paid on the basis of all places, provided that at least two-thirds are utilized for the whole day, or for at least five hours a day. State grants have been structured in this way because the state is concerned primarily with providing help to gainfully employed parents.[14]

The cost of day nurseries is shared among the state, the municipalities, and parents. The state's contribution covers about a quarter of the costs. The parents' fees vary according to income and number of children. But no matter how high their income, parents who have succeeded in getting their children into day nurseries are being subsidized by society. As a rule, the children of unmarried parents enjoy priority of access to day nurseries.[15]

The Child Center—Parents' Views

In the eyes of certain demanding parents the Swedish child center is still far from what it ought to be and perhaps from what it will be— when the importance of the formative early years begins to be realized.

Quantitatively speaking it is inadequate. We have far too few child centers for the intellectually handicapped. These can be children from culturally deprived environments, children whose development has been disrupted, immigrant children from linguistically isolated homes lacking in contacts, or children whose physical handicaps have prevented normal relationships with others. They can also be "ordinary" children who have been emotionally disrupted by continual, wearing friction with frustrated mothers who are incapable of meeting the children's needs for stimulation and outside contacts.

Swedish child centers are also qualitatively imperfect. This is in spite of the fact that we have more well-trained preschool teachers every year and are one of the few countries where supervision and pedagogics have been officially accorded equal importance. The latter means that all those publicly employed to look after children should, in principle, have some form of pedagogic training, plus practical experience in caring for children in a family or institution. In practice this is not yet possible. Children under five years attending a day nursery are usually looked after by children's nurses who have had only a short period of training and have no great pedagogic insight. Materially speaking the children are very well looked after.

Praiseworthy attempts have been made to reduce the difference in pedagogic standards between the day nurseries and the previously more advanced nursery schools. However, a number of local authorities have not yet understood the need to staff day nurseries with people who can give children more than physically adequate care. They still have an entirely out-of-date view of the day nursery as a sort of social assistance.

The intensified program for the building of day nurseries in recent years has been dictated largely by the requirements of the labor market —the need for "supervision" of children of working parents. The view that day nurseries and nursery schools are needed for the sake of the children themselves, for their development, has begun to gain acceptance only recently—although it has been held by preschool teachers themselves for more than half a century.

Regarding the content of preschool activities, we are still in the experimental stage. The emphasis has been on providing emotional security by offering the children ample opportunity to develop their senses in play. In this we have come further than many other countries. However, we are now beginning to realize how much more can be done to stimulate the children's intellect, imagination, and emotional life, to increase their understanding of others and of the world around them, and to give them an opportunity to identify with adult activities. A great deal remains to be done before the children at all preschools are trained in communicating with others by words, rhythmics, music, and dramatic games.

The absence of male supervisors and teachers is a shortcoming that Swedish child centers share with all other countries—with a few exceptions in Denmark. The consequences to small children of living exclusively in a female society has been demonstrated by, among others, the Scandinavian research scientists Per Tiller and Gustav Jonsson-Kälvesten, who have confirmed that the absence of males has negative effects on boys and girls alike.[16] In spite of our newly acquired insight into this, the responsible authorities have not yet taken any radical measures to bring male staff into the child centers, for instance, by increasing the salaries and career opportunities for well-trained teachers. The absence of males is particularly serious insofar as it is principally the children of single parents, for the most part single mothers, who attend day nurseries, children who have predominantly female contacts' outside the nursery.

As yet the parents' opportunity to cooperate with the staff of the day nursery in the everyday care of the children is very limited. Cooperation with the parents has simply not been considered and stimulated to a sufficient degree by the responsible authorities. The preschool staff has too little resources, too little money and time, too little training in adult psychology to be able to promote active participation by parents in the work of the day nursery. The parents have failed to band together and exercise pressure to obtain legislation that would permit them to devote more time to the children in the early years.

In spite of the criticism that can be made of the Swedish nursery school in its present form, the majority of modern parents seem to want to retain and develop it. Many people would like to see the nursery school expanded to include three year olds, and view with some suspicion the proposed introduction of a compulsory nursery school for six year olds. Will this not mean that all qualified preschool teachers are assigned to the six year olds? May there not be less chance then of achieving pedagogic stimulation in new, untested forms for the many younger children who risk being retarded in an emotionally or culturally deprived home environment?

Some people are also afraid that preschool teaching will become more rigid and structured, that the nursery school will be adapted to the school, instead of radical reforms being made in teaching in the early years of compulsory school and in the training of teachers for the junior level.

The current revolt of young people the world over shows that our present society and its ideals have not won the approval of the coming generation. But what do they or we who are parents of the next generation want instead? The only thing we know for certain is that the small individual being shaped today will be faced, as an adult, with greater demands for empathy, solution of conflict, flexibility, and the suppression of egoistic motives in a shrunken, unfrontiered world of dwindling resources and growing needs. It would seem that we should consider the consequences of this fact in our present planning for children.[17]

PRINCIPLES OF SWEDISH SOCIAL POLICY AND ATTITUDES TOWARD WOMEN AND CHILDREN

General

The Swedish view is that everyone has the *right* to help from the community when he needs it. Voluntary efforts can only supplement those of the community, and their value lies precisely in offering something beyond the routine. Charities must never be an excuse for public neglect or delay efforts financed by taxation. Only public services based on public taxation can be the legislated right of all citizens.

The expansion of the social-service system has afforded an experience that would previously have appeared paradoxical, namely, that the individual's demand for security and social care rises with his affluence. Once a high standard of living has been achieved, there is more reason to aim at greater security rather than further increases in the standard. The man who has much to lose has correspondingly much to protect, and he wants a guarantee that he will not find himself in financial difficulties that may devastate his home and family life.

The general view of social security has altered. Previously such measures were regarded purely as a burden of central and local government on the economy. Today social policy is regarded as an important positive factor in that it offers the individual not only greater security but also a greater chance of making an active contribution in production and the life of the community. In other words social policy pays off in the national economy.[18]

Attitudes toward Women in Sweden

In Western culture the traditional view has been that man and woman were created different, and that this difference must be reflected in their whole mode of life, both in family and society. Traditionally a woman is regarded as better suited to taking personal and practical care of children, while a man's job has been to earn the family's living and represent it vis-à-vis society. If this fundamental distinction is abolished we can, according to Per Holmberg, a Swedish economist, greatly improve our standard of living.

Per Holmberg has figured out that Sweden could increase her national product and income by about half if women were as productive and gainfully employed as men. He comes to the conclusion that we are paying an exceedingly high price for the separate life roles of men and women—in fact foregoing almost one-third of the material standard of living that could be ours if those roles were the same. The expression "life role" (or sex role) is meant in a social, not a sexual, sense.

Per Holmberg is not the only person in Sweden who would like to abolish the radical distinction between the life roles of the sexes. On the contrary, it is nowadays most unusual for any Swede in a responsible position to defend it. (In principle all Swedish political parties agree that there should be equality between the sexes in all areas of life in the future society.)

At a conference held in October 1967, five highly influential executives answered a number of questions about women and the labor market. The degree of agreement that reigned among these five distinguished men—that everyone, regardless of sex, has the same right to be gainfully employed, that women must be regarded as every bit as valuable a part of the labor market as men, and that men must have the same practical and emotional responsibility for children as women—was quite touching to behold. They also thought it self-evident that the father as well as the mother of small children should be entitled to take a part-time job or forgo the advantages of overtime.[19]

Elsewhere a young academic has given a personal account of what it is like in practice to achieve equality of sex roles:

Nothing irritates me more than to be praised by well-meaning persons for "helping" my wife look after our children. Patiently, I try to explain that it isn't a question of "lending a hand," and that one could just as well praise my wife for "helping" me. We devote ourselves to the children equally and cannot possibly find any sensible reason why we shouldn't. To our ears it sounds almost an insult that anyone should assume that my wife looks after the children alone: as if I were negligent, didn't like my own children, weren't competent, or something equally offensive.[20]

Under Swedish law man and woman are formally each other's equals, even within the family. The marriage law of 1921 made them so. But many people in Sweden today still support the traditional and romantic idea of male and female natures and sex roles. Such ideas are incessantly nourished by the weekly press, by advertising, by books for young people, and by television. Nevertheless, other "ideologies" concerning the life roles of the sexes have taken root.[21]

Attitudes toward the Family

The number of families in which both spouses are gainfully employed is continually increasing. Thus Sweden is approaching a situation in which it is normal for a family to have two members earning incomes.

The increasing part played by married women in the labor market has resulted in entirely new demands on family policy. This policy must be adopted to the changed roles of the sexes and must be designed to promote equality in the labor market and in the division of labor at home. Increased prominence will be given to care of the children and the cost thereof.

Up to now the Swedish government has adhered to the principle of evening out living standards, as far as possible, between the period when the family's needs are greatest and the period when its maintenance burdens are lightest. Without this kind of evening out, living standards are bound to fall considerably during the years when the children need care and attention. Therefore, additional support over and above the basic child allowance is necessary during the period when the children most need attention. The political program for women adopted in 1964 laid down the following guidelines for family policy: (1) social insurance comparable to that available to the working population should be provided for the spouse looking after the children; (2) payment should be made for such child care; (3) the child's direct consumption costs should be provided for.

Since it is to be expected that women will continue to play a more active part than men in looking after the children for a long time to come, a family policy based on these principles would be of considerable importance to them. Social insurance during the period when children require most attention would also encourage parents to resume working when the children are old enough to take care of themselves. In this way child care and outside employment combined would give parents the same old-age pension as persons with full-time employment are guaranteed through the general supplementary pensions scheme alone.[22]

Anyone who wants to tear down the obstacles to women working outside the home must not only try to reduce the difficulties that women confront in the labor market and arrange for the supervision of small children, but must also influence family members to rethink their whole way of living so that they will share common tasks and not thoughtlessly leave all household chores to mother. It seems ill-advised to adapt measures intended to facilitate women's access to the labor market in such a way that a woman still has two roles—notwithstanding the fact that women do, in reality, have two roles—two jobs to do. An official study has shown that the Swedish man does not do his share of housework. Those fathers who do least in the house are those who have the most children. When the woman goes out to work the man's contribution increases slightly. When the woman works as a housewife 12 percent of the men help with the cooking. If she goes out to work 29 percent of the men "lend a hand." Nor does the father spend much time looking after the children, though he commonly has no objection to playing with them.

The Swedish family is certainly not going to revise its whole attitude toward life overnight, and, of course, both men and children get used to evading their share of housework during the years when so many women interrupt their working life to take charge of these duties. For a long time to come women will probably have to live with their double role. It is to be hoped, however, that not too many of them will accept it or rest content with a martyr's halo.[23]

Some Alternatives—New Family Patterns

Marriage and family structure are being debated and questioned. The isolated nuclear family is criticized. New types of families and interpersonal relationships are discussed and experimented with: modern group families as well as one-parent families, common-law marriages, trial marriages, group marriages, sometimes also involving group sex. Communal group living as well as househusband families contribute to the shifting of the social roles of husband and wife.

In most cases such types of families should be regarded as serious attempts to create better emotional conditions for both children and adults alike. The psychological consequences of the general reevaluation of sex roles may be seen in some of the advanced young couples. Some of them stay childless, feeling that it is unethical to give birth to more children in this overpopulated world. Others have one or two children of their own and then adopt children from deprived areas in other countries. Consequently the value of womanhood is not tied to childbearing and the value of manhood to biological reproduction as in old times.[24]

Men

The new tasks of men and women in society and in the home do not only mean changing roles for women but also the emancipation of men! The greatest disadvantage with the male sex role is that the man has too small a share in the children's upbringing. The ability to show affection and to establish contact with children has not been encouraged in the man. From infancy both boys and girls need good contacts with adults of both sexes. Studies reveal a common trait in the histories of children and youths with various kinds of behavioral disturbances: they have poor or no contact with the father or any other grown-up male.[25]

The sociologists and psychologists draw particular attention to the identification problems of the boys. Already at the age of three the child needs to identify himself or herself with somebody of the same sex. This process is easier for the girls because they have constant contact with women. It is more difficult for the boys. In modern society they grow up almost wholly in a female world. At home they are as a rule taken care of by their mothers. During the early school years all their teachers are women. There is a risk that boys will develop by means of TV, comic strips, and other mass media a false and exaggerated picture of what it

means to be a man. The men are tough and hard-boiled Wild West he-
roes, agents, supermen, soldiers. The boys compensate for their lack of
contact with kind, everyday men by looking upon mass media men as
their ideals. It should be possible to counteract these problems. Men
should have just as much contact as women with their children from the
time they are born, and both men and women should be child nurses,
kindergarten teachers, and infant school teachers.

A few years back we had a rather intense discussion in Sweden on
whether mothers of small children should work outside the home. As a
result of the new view of life roles, the problem will be instead should
the *parents* of infants be employed. One solution is that parents work
part-time and take turns looking after the child. Many young families
with flexible working hours, for example, undergraduates, now practice
this arrangement in Sweden.

The new role of the man implies that he must reduce his contribution
to work—and maybe also to politics—during the period when he has
small children. This is what the woman alone has always had to do.
From the point of view of the national economy we can manage this loss
in production if we can instead stimulate women to make increased contributions.[26]

POLICIES FOR CHANGE

We look upon the emancipation of the man as important for the development of our children and for equality between the sexes. The different parties have drawn up programs demanding that men and women
should have the same roles. The big trade-union organizations have
prepared their own programs that will make it possible for men to share
child care with the women. The trade-union organizations and the employers' organization also have a joint body that works for equality between the sexes in accordance with this principle. The same views are to
be found in the report on the status of women in Sweden that the Swedish government submitted to the United Nations in 1968.[27]

The Future: Universal Preschool Care

At present many children cannot obtain places in child centers—
whether both or only one of their parents is at work during the
day. The need arises to make it possible on a much larger scale for
children to attend preschool for part of the day over several years. The
Swedish Central Organization of Salaried Employees has demanded
such a reform. It wants to ensure that all local authorities—with strong
support from the state—offer all children the possibility of attending
preschool for at least two years. This question is now being considered
by the Royal Commission on Child Care. The directive given to the

Royal Commission emphasizes that the aim of educational planning in recent years has been to offer a good school education regardless of the district of residence, financial status of the parents, and other circumstances. It has also been stressed that this process of democratization should be broadened to cover circumstances influencing the initial position of the child on starting school.

This will make great demands on the preschool system and will probably require some change in its aims and in the structure of its activities. This, however, is something that the Commission must first consider. Even if the question of establishing a fixed curriculum does not arise, the Commission's directive suggests that there is reason to compile more concrete recommendations on the nature and structure of activities. Studies on the effect of attending a preschool suggest that it is often relatively slight in the case of children from relatively well-off families who themselves make an effort to promote the child's development —largely families with a good education. In the case, however, of children from less adequate environments, the effect of preschool attendance is marked. It is striking how the consequences of these findings have been ignored.

To begin with it is obvious that children from different environments get a very different start in life—a situation that could be improved by a real investment in preschools. Secondly, these studies suggest that children generally have a developmental potential that is not realized either by the homes or by our present type of preschools. We suspect that the adults determining the conditions under which our children grow up—by virtue of their position as parents, teachers, or public officials—do not really know what is best for the children. In many counties the preschool is a more conservative institution than the school. This is natural enough, since the younger the child, the greater the degree of control that adults can exercise. Also, preschools in Sweden— and in other countries—are not subject to the same reformatory zeal and interest on the part of the informed opinion. Nor does legislation provide for the same supervision as in the case of schools.

However, the view that children need both the preschool and the home is gaining acceptance in Sweden. It is unreasonable to demand that the parents should meet all the child's needs, still less that the mother should accept responsibility for the child's upbringing to the extent she does now. This responsibility must be shared by *both* parents, both of whom need outside support.[28]

NOTES

This chapter is a collage of pieces on child care by various Swedish writers. The kind assistance of Ingrid Arvidsson of the Royal Swedish Embassy, Washington, D.C., Marianne Kärre of the Swedish Commission on Child Care, and Mats Hellström, chairman of that commission and Member of Parliament, is gratefully acknowledged.

After this book went to press, the Swedish Royal Commission on Preschool Education and Day Care, appointed in 1968, released its final 1200-page report. Its recommendations for the expansion of preschool education will be considered by Parliament in 1973 and a new preschool law, if passed, will go into effect in 1975.

1. Marianne Kärre, "Child Supervision in Sweden," *Hertha,* no. 5, p. 60.

2. Åke Fors, *Social Rights in Sweden: Social Policy and How It Works,* trans. Keith Bradfield (Stockholm: The Swedish Institute, 1969), p. 16.

3. Anna-Greta Leijon, *Swedish Women—Swedish Men* (Stockholm: The Swedish Institute, 1968).

4. Astrid Wester, *The Swedish Child: A Survey of the Legal, Economic, Educational, Medical and Social Situation of Children and Young People in Sweden* (Stockholm: The Swedish Institute, 1970), p. 11.

5. *Ibid.,* p. 12.

6. Fors, *op. cit.,* pp. 16–17.

7. Wester, *op. cit.,* pp. 20–21.

8. Siv Thorsell, "For Children's Minds—Not Just to Mind the Children," in *Social Rights in Sweden: Before School Starts* (Stockholm: The Swedish Institute, 1969), p. 9.

9. Kärre, *op. cit.,* pp. 60, 61.

10. Thorsell, *op. cit.,* p. 6.

11. Kärre, *op. cit.,* p. 60.

12. Thorsell, *op. cit.,* p. 9.

13. *Ibid.,* pp. 9, 10.

14. *Ibid.,* pp. 10, 11, 12.

15. Leijon, *op. cit.,* p. 92.

16. Cf. Per O. Tiller, "Father Absence and Personality Development of Children in Sailor Families, A Preliminary Research Report, Pt. 2," In Nels Anderson, ed., *Studies of the Family* (Goltingen, West Germany: Vandenhoeck and Ruprecht, 1957) 2:115–137; Gustav Jonsson-Kälvesten, *Delinquent Boys: Their Parents and Grandparents* (Copenhagen: Munksgaard, 1967); John Nash, "The Father in Contemporary Culture and Current Psychological Literature," *Child Development* 36, no. 1 (March 1965): 261–297.

17. Marianne Kärre, "The Child Centre—As Seen by a Parent," in *Social Rights in Sweden: Before School Starts* (Stockholm: The Swedish Institute, 1969), pp. 16–22.

18. Fors, *op. cit.,* pp. 3, 4, 10.

19. Leijon, *op. cit,* pp. 31, 32, 33.

20. *Ibid.,* p. 37.

21. *Ibid.,* pp. 39, 40, 112.

22. Maj-Britt Sandlund, *The Status of Women in Sweden: Report to the United Nations, 1968* (Stockholm: The Swedish Institute, 1968), pp. 56, 57.

23. Leijon, *op. cit.,* pp. 100, 101, 102.

24. *Ibid.,* pp. 38, 40.

25. *Ibid.,* p. 41.

26. Olaf Palme, "The Emancipation of Man," address by the Swedish Prime Minister at the Women's National Democratic Club, Washington, D.C., June 8, 1970, pp. 7, 8, 9.

27. *Ibid.,* p. 9.

28. Thorsell, *op. cit.,* pp. 14, 15; for further information on Swedish child-care policies see Oscar E. Stranberg, ed. *Social Benefits in Sweden, 1970* (Stockholm: The Swedish Institute, 1970); 1968 Royal Commission on Child Care, first report, *Innehåll och metoder i förskoverksamheten, 1970;* Richard J. Passantino, "Swedish Preschools: Environments of Sensitivity," *Childhood Education* 47, no. 8 (May 1971): 406–411; Steven Kelman, "A Nonchalant Revolution: Sweden's Liberated Men and Women," *The New Republic* 164, no. 11 (March 13, 1971): 21–23.

11

CHILDREN'S DAY CARE
IN FINLAND

IRJA ESKOLA

On a spring day in 1968, the eve of Mother's Day, a gay and colorful parade appeared in Helsinki. Children, baby carriages, mothers, and fathers marched up to the House of Parliament with balloons and placards, some of which read: "A child is not a little thing." Speaking to the members of Parliament, they demanded that more attention be paid by society to the expansion and improvement of day-care facilities for children.

Undeniably the debate over sexual roles was one of the central issues of the 1960s in Finland. A reform movement arose, with the acknowledged goal of changing the prevailing role structure and division of labor between the sexes, making it both more just and better adapted to the needs of society. At the same time it was observed that the position of the child was intrinsically involved in changes in the parents' roles. The debate over the child's rights, over the child's own culture, and in particular over children's day care has been especially lively during the last two or three years. By means of the type of demonstration described above, as well as by means of the public media, Finnish parents have been aroused to make the 1970s the decade of the child.

In addition to this public debate, the problem of day care has been a very concrete one in many families. While in 1955 only about 33 percent of mothers with preschool children worked outside the home, in 1965 the proportion was 40 percent. The number of day-care services has not increased at anything like the same rate. Public day-care institutions can cope with only a small proportion of even the children of mothers with full-time jobs.

Since the proportion of mothers of preschool children working outside the home is expected to increase steadily, and since it is recognized that from the point of view of mental and social development it is beneficial for the child to participate in supervised group activity before beginning school, the organization of adequate day care has become one of the biggest problems facing Finnish social policy during the 1970s. Both in extent and in cost it is of the same order of magnitude as was the establishment of the public education system in its time.

The Position of the Child in Finland

MATERNITY AND CHILD CARE

Finland's infant mortality rate is among the lowest in the world; it occupies fifth place in the UN statistics. A crucial factor in the sharp reduction of infant mortality is the network of prenatal and child-care clinics, established by law and reaching practically every pregnant woman and newborn child in the country. By 1966, 98 percent of all new mothers were registered with the clinics. The present objective of the clinics is to have every expectant mother visit the nurse-midwife at least once a month and the doctor at least three times during her pregnancy. In addition she receives a medical checkup within twelve weeks after delivery.

In 1966, 94 percent of all live-born children under one year of age were registered with the clinics. The purpose of these institutions is to practice preventive health care; the clinic functions as a health care center, specifically for healthy children from two weeks old to school age. Every child should visit the pediatrician at the clinic at least three times during its first year and once a year subsequently. In addition the clinic gives practical guidance on the feeding, dressing, and other care of children. Both the maternity and the child-care clinics are absolutely free of charge.

MATERNITY LEAVE OF ABSENCE

According to one recent research study, in 1967, 90 percent of Helsinki mothers expecting their first child and 54 percent of mothers expecting their second child were gainfully employed outside the home.

Current labor legislation guarantees women the right to maternity leave during the period for which they receive maternity benefits under the National Sickness Insurance Plan, i.e., seventy-two working days. The earliest the new mother may return to work is six weeks after delivery. The employer is not legally required to pay the mother's salary during this period of leave. It has, however, been agreed in many contracts that certain employers, such as the state and local authorities and some private employers, will pay a full salary during maternity leave. If a mother receives a maternity leave salary equal to or more than her health insurance maternity benefit, she is not entitled to the benefit for the same period; this is then transferred to the employer. A woman who does not receive a salary for her period of leave has her loss of income compensated for by the health insurance maternity benefit, which covers seventy-two working days and whose size is determined by the mother's income; it is equivalent, on the average, to 45 percent of the mother's salary. Of this sum one-third is paid before the expected time of delivery, one-third immediately following the birth of the child, and one-third following the postdelivery checkup.

In addition to these benefits, every mother whose pregnancy has lasted at least 180 days receives from the municipality a maternity benefit, in the form either of money or of the so-called mother's package, which contains items needed for child care: baby clothes, diapers, soap, lotion, and the like.

FINANCIAL ASSISTANCE TO FAMILIES WITH CHILDREN

According to current child-allowance legislation, every child below sixteen years of age living in Finland receives benefits from public funds for his or her subsistence and education. The size of these benefits is scaled according to the number of children in the family. In 1971 the benefit for the first child was $62 annually; the benefit for the first and second child together, $136; and that for three children, a total of $224 annually. Thereafter for every following child the benefit was $88 annually (the average Finnish annual wage or salary is approximately $1,600.). This benefit is paid to every family, regardless of income, and is usually payable to the mother.

In addition special child benefits are paid to families in particularly difficult circumstances in order to ensure the child's welfare. In 1967 this extra benefit was paid to some 6 percent of all families with children. Finally certain measures designed to equalize family expenditures, such as family allowances, housing allowances to large families, and tax deductions, also help improve the position of the child.

COMPULSORY EDUCATION

In 1971 a law concerning the basic school system was passed in Finland, according to which all children between the ages of seven and fifteen will attend the new nine-year basic school. School attendance begins in the autumn of the year in which the child reaches his seventh birthday. With the exception of certain experimental projects, no preschool education in Finland is organizationally integrated into the school system; supervision and direction of preschool children's activity take place within the nursery school system, insofar as space permits, but in 1968 only 10 percent of all six year olds could be placed in nursery schools. In 1970 the government appointed a kindergarten committee, whose task is to define the objectives and curriculum of the kindergarten and to draw up a recommendation for legislation on the subject.

The Need for Day Care

The criteria for determining the need for day care have been the subject of considerable debate. Factors affecting the extent of day-care services are the number of children of day-care age, the extent and time span of the parents' employment, and the possibility of arranging care for the children at home by someone other than the parents. The question of what type of day-care services is needed is affected in turn by the special needs of the children using these services and by the wishes of their parents.

The birth rate has declined in Finland more rapidly than was expected. Emigration from the country has also been greater than expected, and the emigration of families with young children contributes to the declining size of the youngest age groups. For these two reasons it has been particularly difficult to draw up reliable forecasts concerning the size of the population group from birth to six years.

EMPLOYMENT OF MOTHERS

The figures on the participation of Finnish women in the labor market are among the highest in western Europe. In 1955, 35 percent of mothers with preschool children worked outside the home; by 1965 the proportion had risen to 40 percent; and in 1967, 51 percent of these mothers were gainfully employed outside the home.

The urgency of the day-care problem is also indicated by the fact that a number of municipalities, principally cities and towns, have made their own local survey of the day-care situation and needs in their area to help them in drawing up an overall day-care program. On the basis of these individual studies, we can conclude that about half of all mothers with preschool children work outside the home. The lowest relative proportion—38 percent—was found in northern Finland, in the Rovaniemi rural district, and the largest—64 percent—in the south, in the capital city of Helsinki. According to a sample survey in Helsinki in 1970, 45 percent of Helsinki's mothers with preschool children were employed full-time outside their home, 8 percent were employed part-time outside their home, 7 percent were gainfully employed in their home, 36 percent were not employed, and 3 percent fell into other categories (study, maternity leave, random work, etc.).

The same study showed that 74 percent of those mothers who had one child below seven were gainfully employed, while 44 percent of those with three such children were employed. Sixty percent of mothers with primary school degrees, 66 percent with vocational school degrees, and 76 percent with higher academic degrees were working. It is thus evident that the higher the educational level of the mother and the fewer the children of preschool age, the more likely it is that the mother will be working. As the educational level of the population in general rises (and this rise is especially strong in the case of women; at present there is a female majority in the universities of Finland, achieved in 1964–1965), and as on the other hand the number of children in families gradually decreases with the declining birth rate, it is highly probable that more and more mothers will enter the labor market. The proportion of working mothers increased in Helsinki during the five-year period 1965–1970 by 9 percent. If the increase continues at the rate of 1.8 percent annually, the proportion of Helsinki mothers with children below seven working outside the home will be 62.5 percent in 1975 and 71.5 percent in 1980.

The Basis for Day Care

At present there is no legislation in Finland concerned with children's day care as a whole, but such a law is currently being prepared. On November 20, 1970 the government appointed a committee whose task was to prepare an overall program for the organization of children's day care for the country as a whole, including the necessary legislative and other recommendations.

PRESENT LEGISLATION

Day care sponsored by the public authorities is now based on the Child-Welfare Act of 1936. This law does not mention the concept of day care; rather, it requires the municipalities to establish and maintain institutions that support and complement home care and to assist similar institutions established and maintained by private individuals. This requirement has not been interpreted to obligate the local authorities to organize day-care services in their area.

In the act concerning state assistance to nursery schools, dating from 1927, it is assumed that nursery schools serve only children between the ages of three and seven. According to the 1936 Code of Regulations on the same subject, "The working day of the nursery school is at least four hours long."

In practice this has meant that the actual daily period of activity of the nursery schools is four hours. Along with this part-time nursery school activity, it has been found necessary to create other forms, with the specific aim of making available full-time day care. Such forms are the all-day sections of the nursery schools, the infant creches, and the extended day-care centers that include both a creche and a nursery school. According to the guidelines laid down by the Board of Social Welfare, all-day nursery schools may be established in case of need, primarily to serve those children "whose mothers are compelled, in order to earn a living, to seek gainful employment outside the home." Creches and extended day-care centers are based on the same principle.

The Nursery School Assistance Act also contains specific regulations concerning the supervision and control of the institutions. In order to obtain state assistance, they must satisfy certain requirements as to room space, personnel, size of groups, and so forth. In addition certain requirements related to health care are specified in the Health Care Act.

THE PURPOSE AND PRINCIPLES OF DAY CARE

As can be seen from the above, the official attitude is that part-time nursery schools, complementing home care, are primary, while all-day care for the children of mothers compelled to work for financial reasons is a necessary evil. This point of view has an adverse effect on the work of the full-time sections of the nursery schools in particular. In recent years there have been considerably more applications to the public day-care institutions than it has been possible to accept; therefore, those accepted for the full-time sections have generally been the children of single parents or of families in particularly difficult social or economic circumstances. Thus the full-time group of children, most of whom are accepted on social grounds and come from disadvantaged home environments, is characterized by a certain homogeneity. This in itself has an adverse effect on the children's development. The part-time children, with a more heterogeneous and in general a more stimulus-rich social background, are in an advantaged position compared to the full-time children. This difference leads to a certain stereotyped attitude toward the two groups, and a differential value is attached to them. It should, however, be observed that since in Finland the part-time and full-time children are placed in the same schools and also in the same activity groups, some of the disadvantages stemming from the admission criteria can be eliminated.

According to the instructions from the National Board of Social Welfare to the directors of nursery schools in 1969, when filling places in the part-time sections first choice should be given, whenever possible, to six-year-old children in order to prepare them for the group activity characteristic of regular schools and in order to develop their learning readiness.

Thus the outcome of the present situation is that full-time public day care is stamped by an attitude of social welfare, and part-time care by a concern with pedagogic needs. The new legislation now being proposed is intended to eliminate this division and to offer children's day care to parents as a social service, which they may buy according to their needs.

Day-Care Services

DAY CARE IN INSTITUTIONS

At present the only form of children's day care organized by society is that taking place in institutions. Along with these municipal services, a broad private sector has also arisen.

History

The beginnings of day care in Finland date back to the 1850s. The first nursery school in Helsinki, intended for the children of wealthy families, was founded in 1883. Five years later another institution, designed this time for children of "the people," was founded, and attached to this institution was the "Ebeneser home," in which nursery school teachers were trained. The first children's creche was founded in 1891 by the Salvation Army.

By 1919 creches existed in nineteen different localities, mainly towns and industrial communities, and could care for some 700 children altogether. In the same year there were nursery schools in twenty-two different places, with a combined capacity of 5,000 children.

It was only in the 1950s and 1960s, however, that the number of institutions began to increase rapidly. Over 65 percent of the public institutions existing today and almost 75 percent of the private ones were founded during these two decades.

Different Types of Institutions

The creche is intended, according to current regulations, for children between the ages of six months and three years; in exceptional cases younger children may also be accepted if, because of the parents' employment or some other reason, they would be left during the day without sufficient care. Creches are generally open throughout the year; their daily hours are usually from 6:30 A.M. to 5 P.M. The creche may be a separate institution, or it may be part of the so-called extended creche, which also includes a nursery school. Creches do not at present receive government assistance.

The nursery school is a day-care institution for children between three and seven years of age. In order to qualify for government aid, a nursery school in a city or corresponding urban area must have at least twenty-five children, elsewhere at least fifteen children. The school may have a part-time and a full-time section, or a number of sections of various types. The maximum number of children in one institution, however, is 100. The part-time section is open for four hours, the full-time section for eight to ten hours. The maximum group of children per teacher is twenty-five; due to the shortage of places, this is in fact the general practice. In the all-day sections two teachers share the working day. The annual working period of the part-time sections (a minimum of 190 days) corresponds in practice to the primary school term. The services of public creches and nursery schools are free of charge; one condition of state aid is that the municipalities may charge only the real cost of meals.

Statistics Pertaining to the Present Situation

Tables 20–1 and 20–2 give information on Finnish day-care institutions in September 1969. The total number of day-care institutions at that time was 661. This figure does not include special nursery schools or day homes for school children. There are also a large number of play schools and day clubs that are open only a few hours a day or a few days a week. The total number of places was about 32,000, of which one-half were full-time and one-half part-time places. Three-quarters of the places were in public institutions and one-quarter in private institutions.

The total number of children of preschool age in Finland in 1969 was 538,480. Thus only some 6 percent of all children could be placed in these institutions. Since, as we have mentioned, some 50 percent of the mothers of these children work outside the home, only 12 percent of the children of working mothers received day care in these institutions.

TABLE 20–1

Number of Day-Care Institutions in Finland, September 15, 1969

Type of Institution	Public [a]	Private [b]	All Institutions
Infant creche	123	28	151
Extended day-care center (creche + nursery school)	43	109	152
Nursery school	269	89	358
TOTAL	435	226	661

[a] "Public" does not mean governmentally financed but refers to institutions open to all.
[b] "Private" refers to institutions open to members only.

TABLE 20–2

Number of Places in Day-Care Institutions, September 15, 1969

Type of Institution	Public	Private	All Institutions
Infant creche	3,368	831	4,199
Extended day-care center	1,442	3,309	4,751
Nursery schools			
Full-time places	6,037	790	6,827
Part-time places	13,228	3,399	16,627
TOTAL	24,075	8,329	32,404

Buildings

When a public day-care center is set up or renovated, the building plans must be ratified by the National Board of Welfare. The institution may not open until it has received the final approval of the Board. Guidelines and recommendations have been laid down concerning the location and room space of day-care institutions. In choosing the location, such factors as the distance from dwelling areas, traffic connections, and healthfulness must be taken into account. The location chosen should be such that the children will have to cross as few streets as possible; and it should provide good opportunities for outdoor play.

In determining the amount and division of room space, factors to be taken into account include suitability for children's activity, quietness, and safety. The norms laid down by the Welfare Board give exact specifications on the size and equipment of the rooms.

Up to now most day-care centers have been built on their own lot in separate buildings. The mean number of places is seventy-five. This manner of building, however, is both expensive and inexpedient, since it means that the children have to travel a long way from their home to the center. Therefore, new approaches are being considered. For example, the Helsinki city budget plan for 1972–1981 assumes that, because of the real estate shortage, the city day-care institutions will be placed in rented or ownership apartments adjacent to other service spaces. This means that the builder will have to be consulted at a sufficiently early stage in order to ensure appropriate room space for day-care services.

Staff

According to the Welfare Board's directives, the personnel of the nursery school include the director, the teachers, a medical doctor, and the necessary auxiliary staff. The creche staff includes the director, children's nurses, and other staff as needed.

A nursery school teacher's qualifications are also specified. The training consists of a two-year course in a nursery school teacher seminar and a six-week summer practice training period. The entrance qualifications for the course generally include middle school (i.e., the first four years of academic secondary school), but a great majority of those applying at present have completed the full eight years of secondary school. The final selection of students is based on aptitude tests. Practically all nursery school teachers in Finland are women, though in the last few years a few men students have entered the seminars. There are five seminars training nursery school teachers; these are institutions maintained by private associations, but receiving state aid. There has been some talk recently of bringing the training of nursery school teachers under state supervision, in practice under the authority of the National Board of Social Welfare.

Table 20–3 contains the Welfare Board's guidelines specifying the personnel for a normal nursery school. The recommended child and staff composition for the creche with thirty children is six to seven babies six

TABLE 20–3
Nursery School Personnel Specified in Welfare Board Guidelines

Nursery School	Director	Teacher	Cook	Nursery School Assistants	Practice Trainees
25 children, part-time	1	–	1	–	1
25 children, full-time	1	1	1	–	1
50 children: 25 full-time, 25 part-time	1	2	1	–	2
75 children: 25 full-time, 50 part-time	1	3	1	1	1
100 children: 25 full-time, 75 part-time	1	4	1	1	2

months to one year old, one to ten children one to two years old, thirteen to fourteen children two to three years old; and one director, three nurses, one cook, two teacher trainees, and one cleaner.

The qualifications of the creche director include, besides competence as an infant's and children's nurse, sufficient theoretical and practical pedagogic training. The nurses working in creches have generally completed a one-year course of study. The weakness of the present system, from the point of view of creche work, is that the present training is predominantly directed toward hospital work, whereas the person responsible for the care and upbringing of healthy three-year-old children actually needs more pedagogic training than is presently given.

The total teaching and caretaking staff of day-care institutions on September 15, 1969 amounted to slightly more than 3,700, of whom helpers and trainees accounted for about 40 percent. In addition to the teaching staff, there were 1,265 persons employed in housekeeping and other auxiliary jobs. In addition to the day-care institutions proper, there were some 500 workers employed in separate play schools and children's day clubs. Of the total personnel 42.3 percent were without professional training; of these, however, a great majority were teacher assistants who had not yet received the professional training for which they were practicing. Of the actual staff under 4 percent were without any professional training for their work.

Numerical Ratio between Children and Staff

As we have mentioned, in order to qualify for state aid a nursery school must have at least twenty-five children, while the group under a single teacher may have at most that number. Because of the great shortage of places, the group maximum is the usual practice. In creches there are no legal norms concerning the size of the groups, but the Welfare Board has recommended the following:

    6 months–1 year   maximum group size 5–7 children
    1–2 years   maximum group size 6–8 children
    2–3 years   maximum group size 10–15 children

Municipal creches have generally tried to maintain these standards.

Table 20–4 shows the numerical proportion between children and staff in day-care institutions on September 15, 1969.

In the new day-care law now under debate, the size of groups will be reduced, so that the number of children cared for in one group in creches will be eight to twelve, in full-time nursery school sections, twelve to twenty, and in part-time sections, sixteen to twenty-four, depending on the age of the children.

TABLE 20–4

Proportion of Children and Staff in Day-Care Institutions, September 15, 1969

Institution	Per Entire Staff	Per Professionally Trained Staff	Per Helpers and Trainees
Creche	4.2	7.0	9.9
Extended creche	7.7	13.4	18.3
Nursery school	8.7	14.8	20.9

In fact most of the criticism of the present institutions has been directed at the overly large size of the groups. The new act will evidently bring considerable improvement into the situation. Discussion of the size of groups brings up the most central problem of the entire day-care field: the relationship between quantity and quality.

The Cost of Day-Care Institutions

According to law, the state helps finance nursery school programs by covering one-third of the annual expenses that are considered reasonable. Municipal creches and a majority of private day-care institutions are not covered in any way by state assistance.

In the case of a municipal institution the municipality pays those expenses not covered by the state. The parents' contribution is limited to the real cost of the child's food. One of the conditions of state aid in fact is that the parents may not be required to pay any more than the cost of the food; if the family is in difficult circumstances the food payment may also be omitted. In 1969 a total of 8 percent of all children in day-care institutions had been freed of all charges.

The costs of privately maintained institutions that do not receive any state support are usually divided among the municipality, the parents, and the supporting organization.

For all institutions the biggest single expense item is that of salaries and social-welfare payments. The proportion is greatest in municipal institutions, where in 1969 salaries accounted for 73 percent of all expenses, and smallest in private institutions without public support, where salaries were 59 percent. Other large items are housing costs, which varied during the year covered from 11 to 17 percent, and food, from 8 to 15 percent of the total.

Almost 60 percent of the income of institutions receiving state aid was from the municipality and only slightly more than 26 percent from the state. About 11 percent of the income of these institutions came from the parents in the form of food fees. The activity of public day-care institutions is thus supported almost exclusively by state and municipal assistance. Private institutions receive support from the municipality, from the church, and from various organizations. About 20 percent of all institutions are without public support of any kind.

The new Children's Day-Care Act now under consideration will recommend that state aid to nursery schools be increased; under the new system the state would cover 50 percent of all salary expenses. Furthermore, municipal creches and private day-care institutions that satisfy certain norms would also come within the sphere of state aid.

At present state aid is limited exclusively to the cost of running day-care institutions. It has been recommended that this aid be extended to include the initial costs of setting up day-care centers. The 1971 state budget did in fact include a sum designated for the use of municipal creches and private day-care institutions. Public assumption of the costs of day care has been restricted in any case to those arising from institutional as opposed to in-home or family day care.

Administration

In order to qualify for state aid a day-care institution must observe the regulations laid down by the Board of Welfare. These rules specify the general principles of the nursery school's working program, its management, the employment duties and salaries of its personnel, its daily schedule, and its janitorial plan.

Institutions that receive state aid are required to have an executive board, which is responsible for its activity. In the case of municipal day-care institutions the municipal Office of Welfare functions as the executive board. Private institutions that receive state aid have their cwn board, elected by the organization that maintains the institution.

Municipal day-care institutions have no parents' council, nor are the parents represented on the executive board. However, one of the tasks of the nursery school teacher is to maintain contact with the parents by arranging parents' evenings and making home visits. Especially popular are the so-called open house days. In addition to these organized events, the parents may visit the nursery school whenever they wish, either for a short period or for the whole day.

The National Board of Social Welfare supervises and directs day-care activity on behalf of the state; at present there are two day-care supervisors. As mentioned above, the Board ratifies the plans and room-space programs of the institutions, and final approval of the Board is needed before the school may begin its activity. In addition the Board approves the code of regulations of the institution, directs its activity through circulars and memoranda, and makes inspection visits. Training and information meetings are organized for the staff of the nursery schools and for the municipal administrators. The curriculum of the nursery school teacher training courses is also subject to approval by the Welfare Board.

At the local level day-care institutions are supervised and directed by persons employed for that purpose by the municipal Office of Welfare and the local health board. As already noted, the local Welfare Office also functions as the executive board of the day-care institutions.

FAMILY DAY CARE

Of the present forms of children's day care outside the home, the most extensive is so-called family day care. By this is meant the daily care of the child in an outside private home, on a commercial basis, either part of the day or all day. There is no public control over this form of care. Anyone who wants to can take in up to four children for daily care (more than that is considered to constitute an institution) without being required to report his or her activity to any authority.

However, a 1968 amendment to the Child Welfare Act gives the child-welfare authorities the power to interfere in a day-care situation when this is particularly demanded by the circumstances. The local welfare office has the authority to decide how many children and of what age may be cared for simultaneously in one day-care home. When necessary the authorities may also terminate a day-care relationship. The question, however, is how the authorities are to learn about unsatisfactory day-care homes since persons carrying on this form of activity are not required to report to anyone.

It is unlikely, at least in the near future, that institutional day care can be extended sufficiently so as to satisfy the total need for day care. On the other hand family day care has proved in many cases to be a useful supplementary form of care. Since in any case already existing family day-care activity required supervision and direction by the authorities, many municipalities have started, on an experimental basis, to develop out of family day care a supervised and directed form of day-care service.

One Experimental Example

The most extensive of these experiments has been running for about a year in Helsinki. The purpose of this project has been to create the conditions for continued development of directed family day-care activity. The experiment took for itself the following immediate objectives: to register the children within the area of the experiment who are at present in family day care and the homes in which they are cared for; to arrange day-care homes for children in the area; to supervise and direct day-care activity; and to train persons carrying on family day care. The experience resulting from the experiment has been carefully studied.

Special attention has been paid to the organization of care for the children of parents working evening or night shifts. During the year that the experiment has been running, it has expanded to include five parts of the city; 340 day-care homes and 475 children have come within its scope. The latter figure represents 6.6 percent of all children in outside family day care in Helsinki. About 1,000 visits have been paid to day-care homes; attention has been paid to the size of the homes, the play and sleeping space reserved for the children, the possibilities for outdoor play, cleanliness, and so forth. With regard to the day caretaker, attention was paid to age, health, personal cleanliness, relation to the children, sense of responsibility, different skills, and ability to develop.

The first contact between the day caretaker and the parents took place in the presence of the guidance worker; at this time various questions related to family day care were discussed, and a written contract was drawn up, defining the period of care, the fee to be paid (including vacation compensation and other fees), the child's meals, naps, cleanliness, indoor and outdoor play, and the like.

During the experiment a thirty-one-hour family day-care course was held in which 114 persons received certificates of attendance. In the area covered there are large numbers of children and an abundant need for day care. In spite of the great activity of the leaders of the experiment, only about 25 percent of the children in family day care in this area have been reached and brought within the sphere of supervised day care; the other 75 percent are still completely unsupervised. The day-care homes reached so far have evidently been of average or high quality; the worse ones have remained outside the experiment. Among its other results this project has shown that every person carrying out day care should be required by law to report to the municipal authority in charge of this activity. This supervision, however, should not be of a policing nature, but rather it should provide guidance and support. At this stage the city did not interfere at all in the size of the fee, which was paid by the parents directly to the day-care mother. The best form of family day care would be for the persons carrying out the care to be employed by the municipality, which would pay them a salary with all social benefits, and recover part or all of the fee back from the parents. In the case of Helsinki this would mean the employment of some 7,000 new workers. Such a decision obviously requires careful consideration and a great deal of time.

OTHER DAY CARE

Home Care and Its Financial Support (Maids, Mothers' Wages, and Children's Allowances)

As is clear from the above discussion, about one-half of the children of working mothers in Finland are cared for in their own homes. A study in Helsinki in the summer of 1970 showed that 30 percent of the children of working mothers were cared for in their own homes by their grandmothers or other relatives, and 16 percent by hired domestic help.

According to this study, the greater the number of children in the family, the more probable it is that the family will have hired help. A survey of domestic employees in Helsinki in 1969 showed that 66.5 percent of all children in the employers' families were under seven years of age, and the families in question reported that they employed help specifically for the care of the children. An above average number of the employer families belonged to the upper social strata. The survey indicated that the present position of domestic help is an advantageous one in terms of both wages and working conditions. Real earnings are relatively high. The greatest negative factor, according to the domestic workers themselves, is the low prestige of their work. Both workers and employers considered the fact that the help usually lives with the employing family one of the greatest drawbacks of the work.

At the moment it is extremely difficult to obtain a maid or a children's nurse; besides, only a few families are financially able to afford such a luxury over an extended period of time. Grandparents' living with their married children is also becoming less and less common. Thus the possibilities of having the children cared for at home are steadily decreasing.

According to the 1970 Helsinki study, 36 percent of mothers with children under seven were at home. Of these, 40 percent reported that their primary motive for not working was to take care of their children themselves. This was further explained by the financial unprofitability of working outside the home, given the present day-care situation, and by the impossibility of finding a decent day-care place for the children. There has been extensive debate over the way in which financial support might be provided for mothers caring for their own children at home. Two forms of assistance have been discussed: the "mother's wage" and the "child-care allowance."

The term "mother's wage" means the payment, out of public funds, of a certain amount of money to all mothers of young children who do not work outside the home and who do not use the forms of day care provided by society. It has been shown, however, that a general mother's wage could not be set high enough to provide a feasible economic alternative for low-income families; on the contrary, it would primarily benefit the mothers of middle- and upper-class families, who are already able to stay at home if they wish. Thus the mother's wage would have a negligible effect, if any, on the need for day-care places. The concept of the mother's wage has also been criticized for inducing women to leave their jobs, even though it is difficult to return later. This arrangement is also considered to weaken sexual equality. Finally it has been pointed out that this procedure does not absolve society from the necessity of establishing day-care services.

This negative attitude toward the mother's wage, however, is not a universal one. It has been pointed out that every mother is entitled, after all, to a free choice of whether she wants to care for her child herself or have someone else do it, and that both alternatives should receive equal economic support. As a means of making available this freedom of choice, the concept of the so-called home-care benefit or child-care allowance has been introduced. This payment would be paid to every child, regardless of whether both or only one of the parents is employed. The parents could then decide themselves how they want to arrange the care of the child: whether they will use the money to pay for day care outside the home, they will hire someone to care for the child in the home, or one of the parents will stay home to care for the child. It has also been suggested that, in order to pay an allowance of adequate size, it should be paid at first only for children below three years. Also since low-income families have the least freedom of choice between work or home, it has been proposed that the benefit be restricted and scaled according to the income of the parents; in this way the money could be channeled to those who need it the most. The inclusion of such a child-care allowance program in the new Day-Care Act is one of the most difficult problems faced at present. It would be not only an extremely expensive solution, but politically a highly questionable one.

PLAYGROUND AND PARK ACTIVITY

There are two types of outdoor play activity: the playgrounds and the "park mothers."

The purpose of supervised playground activity is, by means of suitable toys and games freely chosen, together with directed group activities and sports, to encourage outdoor play and to draw children away from unsuitable or dangerous playing areas to safe and supervised ones. In this form it hardly satisfies the demands placed on day care proper. In practice, however, these playgrounds also serve working mothers in their day-care needs. The child may spend part of the day in a half-day nursery school or other form of day care and the rest of the day on the playground. In the city of Helsinki, for example, there are over thirty playgrounds, a majority of which are open all year. The city covers the costs entirely; there is no fee. The playground also includes a building in which crafts and indoor play may be pursued. During the summer months the children receive a free meal once a day. In 1969 there was an average of 9,200 visits a day to the year-round Helsinki playgrounds.

"Park-mother" activity is a form of baby-sitting or child care carried out by private persons supervised by the municipal authorities. It provides possibilities of outdoor play mainly for children who have a mother or other caretaker at home. The maximum time a child may be with the park mother is generally four hours a day. According to the regulations, the person in charge of the group must receive every child in person and is responsible for him until his regular caretaker comes to pick him up. She is also responsible for seeing that the children are appropriately dressed, that they have opportunities for play consistent with the prevailing weather and other circumstances, and that children who become sick during the play session are brought home. In 1969 there were 110 such park mothers in Helsinki, each of whom had been assigned her own area. The same year, 523 of Finland's 951 parks or playgrounds had directed activity and employed a total of 880 counselors.

The Present State of Day Care

So far there has been no overall research survey of the quantitative distribution of different forms of day care in Finland. The only exact information we have is that, of all children under seven in the country, 6 percent are cared for in municipal day-care institutions.

As has been mentioned, many local authorities have had surveys of their own areas carried out in recent years. Summing up these separate studies, which were conducted between 1967 and 1970, we can say that 50 percent of all preschool children of working mothers were cared for in their homes by parents, relatives, or household help; that 30 percent were in family day care; that 12 percent were in a day-care institution, and that 8 percent were in another situation or without care.

THE PICTURE DRAWN BY ONE STUDY

Of these research projects the most extensive is the survey carried out in the summer of 1970 in Helsinki. In this project the mothers, fathers, and day caretakers (excluding institutions) of a total of 2,000 children were interviewed.

The results of this survey yield the view of the day care of Finnish children in 1970 shown in Table 20–5. The total number of children under age six in that year in Helsinki was 50,138. The researchers thus concluded that some 34,000 children were cared for at home by the mother, father, or other relative, or by a maid or children's nurse. On the other hand about 16,200 children were cared for outside the home, the prevalent form of such care being family day care.

Of the mothers interviewed, 64 percent were employed outside the home, and 36 percent were at home and without any form of gainful employment. As shown in Table 20–5, the largest single group of children of working mothers, 47 percent, was cared for at home. Of the forms of day care outside the home, the largest was family day care, accounting for 27 percent. The municipal day-care institutions were able to offer a place to only 14 percent of the children.

Each of the mothers was asked why the family had chosen that particular form of day care. The most frequent response was that this had been the only one available. The second most common type of response could be called one of principle: only the mother's care was considered good

TABLE 20–5
Day Care of Children under Six in Helsinki, 1970

Type of Care	All Children		Children of Working Mothers	
	Percentage (N = 1,969)	Estimated Number	Percentage (N = 1,198)	Number
Home	67.9		47.2	
Parents	(51.5)	25,820		
Other relatives	(6.0)	3,010	(30.2)	362
Maid	(9.8)	4,914	(16.1)	193
Alone	(0.6)	301	(0.9)	11
Outside home	31.3		51.5	
Family day care				
Strange family	(14.2)	7,120	(23.3)	279
Relative	(2.6)	1,304	(4.3)	52
Day-care institution				
City	(8.7)	4,362	(14.4)	172
Church, employer, private	(5.8)	2,906	(9.5)	114
Other care	0.8	401	1.2	15
TOTAL	100.0	50,138	99.9	1,198

enough for the child. This response accounted for 17 percent of the total. It is worth noting that only 14 percent of the respondents were able to give a detailed rationale for their response; for example, that the present form of care is beneficial for the health, social, or mental development of the child.

Both the mothers and fathers were asked what in their opinion was the best form of care for children. Of these parents whose children were younger than one and a half years, 66 percent considered care in the child's own home, either by the mother or by a maid, to be best; 22 percent would have placed the child in a day-care institution. On the other hand, of parents with children four years and older, only 38 percent still considered home care the best and 51 percent would prefer an institution.

Care for the Sick Child

The question of the day care of sick children has so far received very little attention. The sick child represents no difficulty when the child is cared for in its own home, either by a relative or by household help, but when the child is normally cared for outside the home, considerable difficulties arise.

If a child is sick she cannot be brought to a day-care institution. For those children who fall sick during the day, there is a sickroom where the child can be isolated and made to rest. The institution immediately contacts the parents and requests them to come pick the child up. In the case of family day care, problems are caused not only by the illness of the child, but also by that of the caretaker, since in this type of care substitutes are more difficult to obtain than in institutions.

In the 1970 Helsinki study parents reported several solutions in the case of the child's illness. These are shown in Table 20–6.

TABLE 20–6
Day Care for Child in the Case of Illness

Type of Care	Percentage of Children ($N = 1,058$)
Mother or father remains home	52.3
Relative cares for the child	21.3
Neighbor or acquaintance cares for the child	6.7
Municipal homemaker visits	4.1
Miscellaneous	6.7
No answer	8.9
TOTAL	100.0

Thus in more than half the cases one or the other of the parents—most often the mother—stays home from work to take care of the child. This absence from work is usually deducted either from the mother's salary or from her vacation. Furthermore, absence causes awkward situations on the job and in general induces a negative attitude toward married women in the labor market.

In the experiment with supervised day care described above, cases of the day caretaker's illness were provided for by means of substitute homes, which took the child in temporarily until the child's own nurse recovered. In order to cope with cases of children's illness, a system of substitute nurses has been developed, which provides for a nurse to be sent to the home to care for the child. It has been planned to use retired hospital nurses for this job.

The legislation concerning municipal household assistance provides the possibility of using municipal homemakers to care for sick children. This is in fact the most common solution in small municipalities, where the need for help in families with children is generally not large. But the bigger the municipality in question, the more difficult it becomes to free workers for this purpose. Thus in Helsinki only 4 percent of the parents interviewed reported that they now solve the problem of caring for sick children in this way.

The alternative that seems to obtain the most support in the debate would extend the municipal homemaker program specifically to provide a sufficient number of homemakers for the care of sick children. Another proposed solution is to broaden the scope of the National Health Insurance Program to include payment of the normal daily compensation to mothers for the period during which they are absent from work due to the child's illness. This would, of course, improve the parents' possibilities of staying home with the child themselves, since the loss of earnings would thus be eliminated; on the other hand this procedure would make the position of mothers with small children in the labor market even more difficult than it is now, since it would mean a greater risk on the part of the employer.

Plans for the Future Development of Child Care

The future of the day-care system in Finland will be determined to a great extent by the new Children's Day-Care Act, which is now being debated. The most urgent problem at present is the need for a large number of new day-care places. The overall organization of day care involves the introduction of a consistent and uniform legal basis for the development and supervision of complementary forms of day care. This legislation should include both institutional day care and that taking place in a family-like environment, in addition to supplementary services such as care for the sick child, rehabilitation and care for children with special needs, and the supervision and direction of indoor and outdoor play and other activity.

According to a preliminary statement from the National Board of Social Welfare, the objective expressed in the proposal for the new act is the creation of a day-care network that will offer every child the day care he or she needs. In order to achieve this objective, day-care needs must be taken into account in various sectors of economic planning. For this reason the municipalities have been required to draw up day-care programs spanning a five-year period, which are then approved by the Board of Social Welfare. The organization of children's day care should in fact be a primary and legally enforced obligation. All forms of day care—municipal and private institutions as well as supervised family care—should be brought within the sphere of state aid. It remains to be seen how the support of home care will be arranged. At present while some forms of day care (municipal institutions) are practically free of charge to their users, others (family care) require the parents to pay the full charge. This situation should be remedied by equalizing the expense to parents of all forms of day care. The new act will probably include a system of day-care fees scaled to the parents' income.

Along with financial questions, another important problem requiring a legislative solution is the equalizing of the quality of the various forms of care. At present municipal institutions are subject to a large number of binding regulations and strict controls, while family day care, which is the most prevalent form, is completely outside any form of control. Both family day care and the private institutional sector should be brought within the scope of legislatively enforced supervision.

Finally the new Children's Day-Care Act is expected to make day-
care planning an integral element of community planning, in order to
make available to the parents a sufficient amount of day-care services at
the right time and in the right place.

REFERENCES

The Status of Women
Haavio-Mannila, Elina. 1970. Suomalainen nainen ja mies, Porvoo 1968 Komiteamie-
tintö 1970: A 8, Naisen asemaa tutkivan komitean mietintö. Helsinki.
Suviranta, Annikki. 1970. Preheenemännän ansiotyön kannattavuus ja motivointi.
Helsinki.
The Position of the Child
Vuornos, Maija. 1971. Lapsen asema Suomessa. Porvoo.
Day Care
Helsingin kaupunginkanslian talous- ja suunnitteluosasto: Lasten päivähoito Helsin-
gissä (unpublished manuscript), 1969.
Helsingin kaupunginkanslian talous- ja suunnitteluosasto: Ohjatun perhepäivä-
hoitokokeilun kokovuosiselostus, Helsinki, 1970 (mimeo.).
Kyllikki Korpi-Liisa Päivikki Ailio. Lasten päivähoitolaitokset syyskuussa 1969 ja
Liisa Suviranta: Lasten leikkialue- ja puistotätitoiminta 1969, Sosiaalinen Aika-
kauskirja 6 1970.
Mannerheimin Lastensuojeluliitto, 1970. Lasten päivähoidon kokonaisohjelma. Hel-
sinki 1970.
Lastensuojelun Keskusliiton julkaisu n:o 46, 1970. Lasten päivähoito. Helsinki.
Mannerheimin Lastensuojeluliitto, 1970. Perhepäivähoito. Helsinki.

12

THE DEVELOPMENT OF THE PROTECTION OF MOTHERS AND CHILDREN IN HUNGARY AFTER 1945

SUSAN FERGE

The connection between social policies relating to mothers and those concerning children needs, I believe, no explanation. The well-being of children depends to a large extent on the conditions in which they are born and reared during their early years. This chapter will deal accordingly with family allowances, child-care institutions for young children, and grants and allowances offered to mothers in Hungary. Some other aspects of Hungary's social policies pertaining to the protection of mothers and children, such as special pricing policies and some grants in kind, will hardly be touched upon. In the case of the main child-care and mother-care institutions, however, I shall try to go beyond a mere description and to show some of the political and sociological implications of the measures in question.

Family Allowances

The 1946 monetary stabilization, which followed a difficult period of war losses and heavy inflation, made it possible to extend our system of family allowances and to eliminate many of the social discriminations in the allowance system of the prewar period. In 1946, every industrial worker or employee included in the compulsory social insurance scheme (some 50 to 60 percent of the population) became eligible for family allowances.

Then family allowances were awarded to families with one or more

TABLE 21–1

*Evolution of Family Allowances for Workers' and
Employees' Families (Monthly Sum of Family Allowances in Forints)*

Number of Children	1946	1947	1948	1951	1953	1959	1965	1966
One	10 ⟩14	18	18	30	—	—	—	—
Two	24 ⟩18	36	40	75	75	75	200	300
Three	42 ⟩22	54	66	135	180	360	360	510
Four	64 ⟩26	72	96	210	260	480	480	680
Five	90 ⟩18	90	130	300	350	600	600	850
Six	108 ⟩18	108	168	405	450	720	720	1020
Seven	126	126	210	525	560	840	840	1190
Eight	144	144	256	660	680	960	960	1360
Nine	162	162	306	810	810	1080	1080	1530
Ten	180	180	356	975	975	1200	1200	1700

NOTE: Since June 1, 1966, the sum of the family allowance for a calendar month in the family of an agricultural cooperative member has been 140 forints after two children and 70 forints after every next one. Single parents have larger allowances and are eligible with only one child.

children (see Table 21–1). The sum was not uniform, but was progressive up to the fifth child and then was regressive again. The total amount awarded to a family was seriously restricted by the postwar economic difficulties: for one child it was 2 to 3 percent of the average wage, and even for five to ten children it corresponded to only 18 to 35 percent of the wage. The different amounts awarded for each new child reflected what was felt to be a preferable family size.

The system was slightly modified in 1947 and 1948. The first modification eliminated the former progressive-regressive tendencies, and the 1948 amendment restored the progressive but not the regressive ones. Family allowances were next raised in 1951 to compensate for rising prices and wages; the progressiveness of the allowances remained unchanged. In spite of the rise in wages and family allowances, the standard of living—or more exactly the real income per capita—decreased by 10 percent from 1951 to 1952.

The next revision related to the so-called family protection law in 1953. The criminal laws then introduced prohibited artificial birth control or any surgical abortions. The former measures were strictly enforced. Primarily because of the cold war, the main and explicit objective of this policy was to increase the number of Hungarian children. Also in accordance with this goal, the state penalized people with no children by introducing the so-called tax for childlessness, abolished the family allowance formerly awarded after one child, just tolerated families with two children (their family allowances remained unchanged), and increased the allowances of families with three or more children. This logic led to extending the family allowance to the members of agricultural producers' cooperatives with four or more children.

The period following 1955–1956 has been a time of economic development and easing of political restraints. The abortion and birth control law and the childlessness tax were abolished in 1955. At that time it was finally realized that in spite of all the equalizing tendencies, the income differentiations between various groups remained significant. Despite the general economic development (the real income increased by more than 60 percent between 1950 and 1956), many families were still very badly off. The statistical surveys showed that although the per capita income level of families with one child was about equal to that of the nation as a whole, the per capita income of families with four to five children was about half of the national average (see Table 21–2). It was also found in 1957–1958 that the family allowances covered only a negligible part of the necessary childrearing expenses.

TABLE 21–2

Per Capita Monthly Income in Families of Workers and Employees by Family Size in 1959
(30 Forints = One U.S. Dollar)

Family Size	Per Capita Income (in Forints)	Percentage of Average Income
Childless families	1,077	134.1
Families with one child	838	104.4
Families with two children	668	83.2
Families with three children	539	67.1
Families with four children	449	55.9
Families with five or more children	406	50.6
AVERAGE	803	100.0

SOURCE: *The Income Situation of Workers and Employees' Households, 1959.* (Budapest: Central Statistical Office [CSO], 1962). Data relate only to married couples or families with one parent and child(ren).

As a consequence of these findings and because the number of births had decreased rapidly once laws prohibiting birth control were abolished,[1] the family-allowance law was amended in 1959 in an effort to improve the situation of large families (those with three or more children). This improvement was extended to families with two children in 1965.

The same natalist policy, which aimed to achieve its goals through economic incentives without administrative intervention, led to the 1966 rise in family allowances. After the 1966 modification the allowance for two children was about 15 percent of the average salary (roughly 1,800 forints; 30 forints=one U.S. dollar) and for five children more than 50 percent.[2] This was an important step forward in comparison with the former much lower rates (2 to 25 percent), and it somewhat lessened the gap between smaller and larger families. At the same time some other measures (like the increase of the lowest wages, the increase in pensions, and so forth) had a similar equalizing effect on the distribution of incomes. The percentage of the population with incomes under 400 forints decreased quite rapidly between 1962 and 1967 (see Table 21–3).

TABLE 21–3

Distribution of the Total Population by Per Capita Income Groups, 1962 and 1967

Monthly Per Capita Income (in Forints)	1962	1967
0–600	31.3%	9.8%
600–800	24.0	14.8
800–1,000	17.9	19.1
1,000–1,200	12.1	18.4
1,200–1,400	6.8	13.8
1,400–1,600	3.4	9.7
1,600–1,800	2.0	5.9
1,800–2,000	1.1	3.6
2,000 and over	1.4	4.9
TOTAL	100.0	100.0
AVERAGE PER CAPITA INCOME	823	1138

SOURCE: For 1962, *The Income Distribution in Hungary* (Budapest: Central Statistical Office, 1967), p. 4; for 1967, by permission of CSO.

NOTE: The average increase in incomes (38 percent) was accompanied by a very slight change in consumer prices (about 3 to 4 percent); thus there was a considerable increase in real incomes too.

Nevertheless, family allowances today still cover only a minor part of the expenses connected with children. Therefore, average per capita income and the distribution of income is largely influenced by the number of dependents; a large number of children considerably lessens or almost eliminates the probability of a family's having high per capita incomes, and the probability of being poor increases. One must add that the statistical indicator of per capita income somewhat exaggerates the differences between smaller and larger families. It is well known that the needs of small children are, as a rule, lower than those of adults and that some expenditures (like rents) are not proportional to the number of persons in the household. This double effect may be discounted by cupational group diverge if the difference in the number of their children increases.[4]

These facts lead again to the conclusion that if we wish to increase the stimulating effect of distributing rewards according to work, we must not only differentiate the wages of various occupations but also raise the level of social provisions, especially that of the family allowances for everyone.[5]

These considerations are summarized in the following excerpt from the report of the Perspective Planning Committee for Labor Force and Living Standards, one of the governmental committees elaborating the conceptual framework of long-term plans:

> *The society's contribution to the sustenance of that part of the population which is unable to work should be significantly increased.* This is the condition for abolishing or considerably decreasing the dispersion of family incomes due to the differential ratio of earners and dependents. . . . The improvement of the living conditions of several hundred thousand people who are especially underprivileged and the *liquidation* of poverty depend largely on these provisions, for in the majority of families the relatively high proportion of dependents unable to work is an important factor in their low standard of living.
>
> The two main elements of the social provisions in cash are the social contribution to the sustenance of young age groups which *cannot yet work* (family allowances, child-care grant, etc.) and the social sustenance of those *no longer able to work*, especially older age groups (pensions). Besides revising the pensions that were fixed long ago, we will maintain the purchasing power of the revised pensions over the long run. By 1985 the *average pension should be two and a half times as much as the present one.* Furthermore, it is planned that in 1985 the society will directly cover, by means of family allowances, about half of the average expenses of the upbringing of children (instead of the present 25 percent or so). In such a case *the total cost of family allowances would be four times as high as the present cost.*

In the course of this planning the question also arose whether it was reasonable to reintroduce a family allowance for families with only one child (perhaps a smaller sum than the average).

The nature and changes of attitudes toward family allowances can be clearly seen. In the beginning (up until about 1953) the significance of family allowances was primarily political: the sum was not significant, but the state declared that in contrast with the previous regime it had such commitments, and also that preference was being given to workers and employees, not peasants. From 1953 to 1960 family allowances were regarded mainly as an aspect of population policy. From the early 1960s family allowances have not only been considered as an aspect of population policy but as a welfare benefit and an economic stimulus, that is, as a means to simultaneously accomplish social, economic, and political ends.

Child-Care Institutions

Child-care centers are the second means by which Hungary provides for the well-being of children and mothers. There are two major types of institutions for young children: creches are for babies up to three years old; nursery schools provide for children from three to six.[6] Their purpose is seemingly the same: to provide board and education to the children under their care. In reality, however, their social function and the motives for establishing them are rather different.[7]

CRECHES

The erection of a network of creches, which was practically nonexistent before the war (with a total capacity of only 1,000 children), quickened after 1949. Now we have spaces for about 10 percent of the children under age three. This development, as officially stated, aimed at making it possible for more and more mothers to continue to work after the end of their maternity leave. Not only was the mass employment of women an economic necessity (the extensive development of industries required great masses of new workers, and the level of income, which was generally low, made it necessary for all able members of the family to work), but from the very beginning the idea of realizing equality between the sexes was an important impetus for the creation of creches. This equality has been and is deemed to be attainable. I think this position justified if the right of women to work and the possibilities for them to work are ensured in many ways. That such considerations played a part in the establishment of creches is evident from the fact that the creches can be used only by working mothers. For a long time creches were nearly free. Now the fees, which are proportionate to income and inversely related to the number of children in the family, cover only a small fraction of the costs of maintenance (100 forints monthly of an average 1,000 forints).

Creches are generally well-equipped modern institutions. They are run and supervised by health authorities and are under the control of the Ministry of Health, which is also responsible for the training of nurses and for establishing the main guidelines for the activity of the creches.

There are basically two types of creches, one belonging to enterprises and the other to neighborhood councils.[8] Because of well-known difficulties connected with the transportation of children, there is now a tendency to give priority to the development of neighborhood creches. In fact their capacity has grown in the last years, while that of enterprise centers has decreased (see Table 21–6). This endeavor is now supported by a 1969 government decree according to which it is in the interest of enterprises not to build their own social institutions but to give their funds to the neighborhood council and sign a contract assuring the use

TABLE 21–6
Infants in Creches (Number at end of Year)

	Number of Creches		Number of Children in Creches	
	Enterprise	Neighborhood	Enterprise	Neighborhood
1960	333	483	13,206	18,764
1963	324	583	12,800	23,753
1966	303	678	12,725	30,384
1969	293	739	9,849	30,465

SOURCE: Hungary, *Statistical Yearbook, 1969.*
NOTE: Besides these creches there are some 3,000 places in seasonal institutions, established mainly in villages for agricultural workers.

of council establishments by the workers of the enterprise. (This applies not only to creches but to all types of social institutions.)

In accordance with their basic function, the creches try to adjust their hours to the working hours of parents. On a national average they are open six days a week, twelve hours a day, from 6 A.M. to 6 P.M., but in some cases they are open up to fourteen hours. However, the child-care problem is not entirely solved for mothers who work the afternoon or night shift. In these cases, if there is no other solution, the enterprise may run the creche on a full-time, six days a week basis. There are about 1,000 creches of this type; this arrangement is favored by the ministry who opposes the late evening transportation of the children.

Creches, as is well known, are rather expensive, partly because of the very high staff-child ratio. Actually there are thirty-nine adults for 100 children, but this ratio is planned to increase to forty-three adults. More than half of the personnel are nurses, 90 percent of whom are trained. (There is one trained nurse for six children.) There are different forms of training. The most usual is a two-year full-time or a three-year part-time course. A secondary school degree is not required, but about 80 percent of the candidates do have it. Recently a special secondary school was created for creche nurses that requires only one year of training after graduation. Several on-the-job courses are also offered. In addition to the care given by the nurses, there is a compulsory daily visit by medical doctors to each creche. The doctors (pediatricians for the most part) discuss the children's diets and health problems with the staff and the parents.

Psychologists and educators are not attached to the creches, but only to the newly created Methodological Center of Creches. Their research findings are then transmitted to the practicing nurses. The creation of this center as well as new forms of training is the result of a shift in the orientation toward creches. Until recently the main function of these institutions was the care and supervision of children. Now, partially as a result of findings in infant psychology, pedagogy, and even sociology, a new interest is developing in the educational role of creches. This certainly does not mean the establishment of elaborate compulsory programs, but more attention is being devoted to the mental and emotional development of the child.

In spite of all the efforts and the progress that have been made to date, we still have to face several problems and dilemmas. Some are "only" material: many more creches are needed—there would be 30 to 40 percent more candidates if adequate space was available—and there is a constant need for staff, partly because of the low wages paid in creches. (These were raised by an average of 20 percent in 1971, but the level is still relatively low.) Some other problems are organizational as well as material. Thus, for example, the optimal size of a creche from the point of view of infant development is smaller than the economically rational size, but the latter sometimes prevails and leads to the creation of creches with as many as sixty children. The main difficulty, however, is that creches, at least in their present form, have some intrinsic problems. Transportation of the babies is only one of them; it can be solved by building neighborhood creches. What is more important is that we still do not know definitely whether creches are only a necessary evil, where the most we can do is to try to avoid unwanted consequences, like the bad effects of institutionalization, or whether they may be transformed into an unequivocally good child-development institution. Recent psychological and sociolinguistic findings suggest that conscious educational activity at this period of early childhood can effectively supplement other efforts to overcome cultural and social disadvantages. The solution to this problem certainly needs further research, experimentation, and funds. Until this problem was solved, however, it seemed useful to look for an arrangement other than the further development of creches. For this reason the so-called child-care grant, discussed later in this chapter, was created.

NURSERY SCHOOLS

The desire to guarantee mothers the right to work played a decisive role in the development of a network of nursery schools, which care for children aged three to six. Like creches nursery schools now accept only children of gainfully employed mothers, and on the whole the organization (opening hours, low fees paid by parents, medical service, and so forth) follows much the same pattern. However, from their inception the educational functions of these institutions were generally recognized, and this recognition influenced the activities of the nurseries and the training of nursery school teachers. (See Table 21–7.)

In fact while nurseries have grown more slowly than creches, they have had important qualitative development. Before the war nurseries were almost exclusively social institutions for the poor, working on a half-day basis, providing no food, located in depressing surroundings, having only a few trained teachers, who, forced to supervise sixty to eighty children, could hardly do more than enforce discipline.[9]

After 1945 a huge national movement radically changed this situation.

TABLE 21–7

Main Data about Nursery Schools

Year	Number of Nurseries	Number of Children in Nurseries	Number of Teachers	Percentage of Children in Nurseries to the Total of the Cohort	Children per Group	Children per Teacher
1938	1,140	112,143	1,598	23.6	70.0	70.4
1950	1,773	106,362	2,428	23.5	—	43.9
1960	2,865	183,766	8,538	33.7	31.5	21.5
1963	3,136	184,345	9,776	41.8	28.0	18.9
1966	3,267	191,991	10,566	49.7	27.5	18.2

SOURCE: *Data Thesaurus of Educational Statistics* (Budapest: Central Statistical Office, 1968).

NOTE: Children per group, 1938, is approximate.

It gradually transformed the kindergartens into the best equipped, best financed child-care institutions of the nation. After long years of research and discussion a central manual was issued in 1957 for nursery teachers. It gives guidelines for instructional and educational activities in nurseries. This manual is creating an organic unit between play, work, and instruction in a much more flexible way than in elementary schools. It allows for more individual initiative on the part of both teachers and children. Research is naturally being continued, and the manual is continually improved and revised. A new problem is the gap between the progressive pedagogic ideas and methods of the nurseries and the more traditional methods of the primary schools.

Nursery schools are under the supervision of educational authorities and the Ministry of Education. More and more they are considered an integral part of the whole educational system, a preparatory institution for schools, having an important sociocultural purpose. This explains the changes in the nursery school teachers' training and the aim of the long-term, fifteen-year plan to expand the nursery school so that it will be available to all five year olds.

The training of teachers was and is a basic problem. After a long time when only ad hoc solutions were possible because of the rapid extension of the nursery school network, colleges were created for nursery teachers in 1959. They require applicants to have completed secondary school and give a diploma after two years. In addition to this full-time training, there are several forms of in-service training courses either for untrained persons or for persons who need to complete or update their former training.

Despite this progress other problems remain that I will mention briefly. First, nurseries are unequally distributed throughout the country: villages, where the sociocultural and educational needs are the greatest, have fewer nurseries than the towns. Even if a village has a nursery, there may be no age division in it because of the small number of children in the community, and this hinders the carrying out of educational programs. Second, nonearning mothers cannot place their children in the nurseries, and this adds another element to the inequality of the income distribution: families that have two parents working naturally have higher per capita incomes (the difference is about 25 to 30 percent) than those with one earner, and they profit more from the state contribution to the maintenance of child institutions. In addition families that have nonworking mothers, as will be discussed later, are generally among the less skilled, less educated portion of the population. It is hoped that some of these problems will be overcome in the next ten years by the expansion of nurseries; for others we have as yet no solution. For example, no nurseries are now planned for small villages that have only a handful of children.

A third set of problems is connected with the training and wages of the staff. Tension exists between the different training institutions; primary school teachers' training colleges are three years and have a higher social status than those of nursery teachers. We are now experimenting with merging these institutions or giving the same three-year status to nursery teachers' colleges. But the final solution has not yet been found.

Some years ago we had vivid discussions about the destiny of child-care institutions. Some argued that all these establishments were created in the interest of women rather than children. These groups now argue that since large increases in the labor force are no longer desirable and the redistribution of income is handled more directly, the development of child-care establishments should be slowed down and state subsidization should be lessened at the expense of the parents. According to another belief that now seems to be growing, this hesitation is justified only in the case of creches, but the improvement of nursery schools and all other child-care programs is a social and cultural duty of the utmost importance. Those who hold this opinion also assert that state subsidization should be generally maintained in the interest of children and mothers even if this solution conflicts with some other values underlying the distribution of income.

Pricing Policies for Children's Goods

Some special pricing policies for children also deserve mention. These include low-cost milk provided in large towns for all children under age six and for mothers with infants under one year and relatively cheap children's clothing, school accessories, and toys. All these are subsidized by the state. The total sum of the allowances thus provided is not great; it is fifty to one hundred forints per child per month for milk (about 10 percent of the average monthly per capita family income), and approximately ten to thirty forints a month for all the other articles taken as a whole. Yet milk is important to children's health, and traditionally the consumption of milk in Hungary has been low. When milk is made available cheaply, it not only eases the situation of those with low incomes but also encourages the children to drink more of it. The reduced prices of the other articles are kept more because of tradition than because of the help they offer. At the time of their introduction they were more important; but even today these discounts are not negligible for low-income families.

Grants and Allowances for Mothers

The majority of Hungary's grants to mothers are provided to women shortly before and after they have their babies and to women with young children. Free prenatal care is offered to every pregnant woman and is in fact used by practically all of them. After the child's birth the parents receive a baby's layette worth 400 forints, which since 1971 has been given in cash. In the case of gainfully employed mothers the layette grant is supplemented by a maternity bonus of 700 forints for the first child and 600 forints for the others. Those mothers who are not earners get a smaller bonus. Both the layette grant and the full maternity bonus are paid only to mothers who have had at least three medical examinations during their pregnancy. If this condition is not fulfilled, the sums are reduced. On the whole about 90 percent of all mothers profit from these provisions.

Gainfully employed mothers get a rather favorable maternity leave of five to six months with full wage. Part of this leave may be used in the last period of pregnancy. After the mother returns to work she has reduced working hours until the end of the breast-feeding period. The arrangement is somewhat different in the case of cooperative members; they don't receive a separate bonus and five months' salary, but a lump sum of about 4,000 forints.

Several minor provisions are also available to mothers and infants. These include free breast milk for mothers who cannot feed their babies, paid and unpaid days off from work when children are ill (the number of paid days was recently increased), and special regulations for pregnant women if they are doing hard or uncomfortable work. During the last two years one can observe, moreover, that various enterprises have taken different initiatives and made special arrangements to favor their female workers. Such differential treatment was made possible by economic reforms, and has been strongly encouraged by a decision of the Party's Central Committee on "the political, economic and social situation of women," adopted in February 1970.[10]

The so-called child-care grant, noted above, was introduced in 1967. It does not exist in other countries, except for Czechoslovakia, which introduced it in the summer of 1970.[11] This provision is for working mothers during the period following the maternity leave, until the child is three years old. (According to the original measure, the age limit was two and a half years, but in 1969 it was extended to three years of age because of the lower age limit of nursery schools.) The sum of the grant is 500 forints per month for producers' cooperative members and 600 per month for others.[12] If there are several children under three (twins or others), the sum is multiplied.

The use of the grant does not break the work relation of the woman; she has the right to return whenever she wants, even before the full period is over, and the employer must reserve her former work place for her. It goes without saying that employers sometimes have great difficulties in coping with this provision, especially in the case of semiskilled or unskilled women or when considerable reorganization occurs in the enterprise during the mother's absence. Nevertheless, the legislation will not be altered; we must, however, look for some new ways of implementing it.

This measure is extremely popular. Sixty to 70 percent of all eligible women use it, and the number is constantly increasing; from 34,000 in December 1967, it rose to 92,000 at the end of 1968 and to 144,000 at the end of 1969, reaching a peak of 174,000 in February 1971. This means that about 10 percent of the whole female labor force is profiting from the grant and that more than one billion forints or nearly 1 percent of the entire state budget is spent on this scheme.

The motives for introducing the child-care grant are clear. First, the development of creches is not economical, and it is not an unambiguously good solution. Second, the possibilities for employing an increasing number of people, especially unskilled women, are limited. Thus this measure is helping to avoid unemployment. Third, everyday experience and the findings of time-budget and other research show that working mothers, especially those with young children, are heavily overburdened and need some relief. Lastly, the number of children born after the abolition of abortion laws was for some years disquietingly low: for 1953–1955 the birth rate was twenty-one to twenty-three per thousand women aged fifteen to forty-nine, and then it gradually decreased to reach its lowest value, 12.9, in 1962. From then on it stabilized around thirteen for four years, and then it slowly began to increase. In 1967, the first year of the grant, it was 14.6; in 1968, 15.1; and in 1969, 15.0. Table 21–8 seems to indicate that this rise in the fertility of employed women was already much higher in 1967 than that of nonemployed women. Thus the intended effects seem to have been achieved, and the first consequences appear to be favorable.

Over the long run, however, the measure may have some other secondary effects that are, at least partly, less advantageous. The problems stem primarily from the fact that the rate of use of the grant is highly differentiated among women belonging to different social groups: it is

TABLE 21–8
Live Births per Thousand Women
Aged Fifteen to Forty-nine

	1963	1966	1967
Agricultural population	50	54	55
Population outside agriculture			
Employed women			
Manual workers	48	49	61
Nonmanual workers	53	55	62
Together	50	52	61
Nonemployed women in families of			
Manual workers	73	74	70
Nonmanual workers	31	25	21
Together	64	63	58
AVERAGE OF THE AGE COHORT	53	55	58

SOURCE: *Child Care Grant,* CSO Periodical Publications, 1969, no. 13.

TABLE 21–9
Rate of Use of the Child-Care Grant among All Eligible
Women by the Educational Attainment of Husbands and Wives

Educational Attainment of Wife	Educational Attainment of Husband			Total
	Primary	Secondary	College	
Primary	73.0	68.9	79.2	73.8
Secondary	64.0	57.3	61.8	61.2
College	38.6	29.8	28.8	30.2
AVERAGE	71.4%	59.6%	49.6%	68.0%

SOURCE: CSO, *Child-Care Grant.*

used by 65 percent of all mothers in agriculture, by 74 percent of those in blue-collar work, and by 54 percent of those in white-collar positions. The variation in the use of the grant is even more marked according to educational level (see Table 21–9).

Women with higher qualifications, better jobs, and higher salaries use the grants much less than do others. A threefold conclusion therefore arises. On the one hand it is obvious that income differences between the strata are growing (those initially better off keep their standard while the others' income falls somewhat). This may be accentuated by the possible occurrence of a differential fertility rate, inversely proportional to the financial and cultural level of the family. In fact the measure is more attractive for women with low skills, and, as we know, the ability to make long-term plans is less developed in these strata than in others.

Thus they may choose a solution such as the birth of a second or third child that seems momentarily advantageous, without taking the long-term consequences into account. Such differences in the fertility rate are certainly not desirable, since they overburden the families and do not assure the harmonious development of children. Finally there is the additional problem of the relations within the family. Sociological research indicates that the whole family atmosphere may be more democratic, that the relationships between both husband and wife and mothers and children may become more humanized if the mother has a social standing of her own. The child-care grant in itself does not mean that the mother loses her former status, but nevertheless an additional difference appears here that may also have further effects.

In addition to the impact of the grant on the whole social structure, some other problems arise with regard to the equality of women. Staying at home for a few years certainly means a break in one's professional development and a lag in the increase of one's wages unless there are special measures to counteract these consequences. In our case no legislation has as yet been developed to this effect, but the trade union's headquarters issued a special recommendation urging enterprises to provide normal wage increases for mothers temporarily at home. We do not have records on the results of this recommendation, but it seems to have been implemented in a good number of cases. This measure, however, does not solve all the professional problems of mothers, and there are more indirect consequences as well that are even more difficult to overcome. The grant means that the woman has the *right* to choose to assume for several years (usually two or three, but in case of several children more) the unique role of mother. Now this right may become, in the eyes of men, a *duty* or *norm*. Some men claim that a woman should be *only* a mother and housewife, whose primary duty is to care for her husband and children. Consequently men should not have to assume domestic responsibilities. Hence the grant may reinvigorate the old attitude toward the traditional division of labor within the family. On the other hand a woman's professional career may be hindered by the existence of the grant even if she herself does not use it; in fact managers often hesitate to employ—especially in responsible positions—young women, for fear of having to find substitutes if they later choose to stay at home for some time. Here again the mother is treated as if she alone was responsible for the birth of a child and should assume alone all the consequences.

In both cases the child-care grant is interpreted in a traditional way and is used as a pretext to maintain a former system of prejudices, a system that is fading away only very slowly. Thus the attitudes concerning equality of the sexes that permeate the social consciousness of both men and women are different from the ideal accepted by the society and the ideology it declares. Formerly the gap between these two levels was not manifest and could not even be discussed. Now it is coming to the surface, causing additional problems. But the greater recognition of the issue may provide a better chance for creating conscious change.

At the same time the grant has another sociologically favorable hidden feature. Women's employment, whether we liked it or not, meant extreme burdens for women, for a centuries-old institution like the division of labor within the family could not change in a day. It is true, though, that by taking notice of it and easing the situation of women, we admit that household work is the task of women and keep her situation unchanged. But the lower social status of women really did not stem from the fact that they did not work. In reality women have always worked very hard, only the work they did has never received social recognition. Now, if we admit that work done in the family (not simply household work, but looking after a little child or children) is socially necessary and deserves payment, we may ease the burdens of mothers without prejudicing the situation of women. The child-care grant carries this possibility as well as the less favorable possibilities. Thus on the whole it may be said that the net effect of this measure is unquestionably good in the short run, and its unwanted and less favorable consequences, since they are known, may be treated and gradually counteracted.

Conclusions

What ultimately characterizes Hungary's social policy for children?

If we look at the overall social policy, a few central values, though not always conscious ones, may be clearly seen. First, the general idea of social equality always plays a primary role in social policy, especially with respect to sex and race. For instance, besides the efforts connected with women, we have devoted much energy to solving the question of Gypsies. By the same token economic differentiation, even if it is accepted now as necessary in this stage of our development, is severely limited by the absence of private ownership and by conscious efforts to avoid having less privileged strata lagging behind. Second is the recognition of the central and socially useful role of *work*, both in general and differentiated according to the variance in its social usefulness. A third value is that a socialist society must be dynamic and future-oriented. Therefore, it must pay much attention to the child as its potential future, not only or mainly as a future resource, but as a human being, whose fullest possible development is the measure of the success of our current efforts and the guarantee of the further development of our society. Taking these different, sometimes conflicting values into account, it is not surprising that the most satisfactory solutions were reached in areas where two or more of the above values coincided. This is why the question of children is generally better solved than that of the old; the question of pregnant women or young mothers better than the question of women in general. Similarly it seems true that the problems of children were most successfully solved when their problems were linked to those of working women and the question of women when it also was a question involving children.

Nevertheless, there is still much to do. In the future we need to pay more attention to the relationship between the pragmatic and the ideological aspects of our social policy. We should learn how to better apply our central values in social practice, how to assure the harmony of the basic principles of broad social goals and immediate needs. This would help to counterbalance some potential problems in the case of policies such as the child-care grant.

NOTES

1. More exactly the number of childless families decreased, the number of those with one or two children grew, and the number of families with three or more children decreased. The typical family became that with two children. The conception of the ideal number of children also followed this pattern.

2. The average salary was about 2,100 forints per month in 1970, so the above percentages are now somewhat modified.

3. This is a combined scale, taking into account the differentiated needs of persons of different ages and sexes and also the relative savings of larger households.

4. The data relate to the last year before the reform but the basic tendency discussed here has not changed considerably since then.

5. It does not follow that family allowances should be proportional to wages, for that would be objectionable on other grounds.

6. Besides these institutions, there are half-day school-homes for primary school children, because Hungarian schools run on a half-day basis. About 14 percent of all primary school children (nearly 200,000) receive meals and are supervised in these centers.

7. All the above institutions are for children who live with their families. There are separate institutions for children cared for by the state. Infants are in full-time creches (with about 5,000 places), and older children are placed partly in boarding schools (about 22,000 places) and partly with foster parents (about 11,000). These institutions have quite a number of financial and pedagogic problems that have attracted special attention in the last few years.

8. Hungarian specialists do not envisage the family day-care solution used in some Scandinavian countries and the United States because of the difficulties of control, supervision, and so forth.

9. It should be mentioned that the interwar state of affairs was a clear deterioration from the mid-nineteenth-century beginnings. At that time the new institution, the kindergarten, had high ideals and high standards. The first school for nursery school teachers, founded in 1837, required higher qualifications than normal primary school teachers' training and for a long time included progressive educational ideas. This information is based mostly on the writings of Alice Hermann, one of the best specialists in the field. Cf. Alice Hermann, *Emberré nevelés* (*The Education of Human Beings*) (Budapest, 1947); *Ertelmi elmaradás, értelmi fejlödés az ó vodds otthonokban* (*Intellectual Lag and Intellectual Development in Kindergartens*), 1967; *Óvodás koru gyermekek tájékozottsága a vilagban* (*The Orientation of Kindergarten Age Children*), 1963.

10. We do not have a complete record of these initiatives, but one comes across

reduced working hours for mothers with three or more children; additional free
days; and mothers being allowed to take days rather than evening or night shifts.

11. The Czechoslovakian clauses are similar but not identical to the Hungarian
system. See M. Pavlova: "Women at Home and at Work," *Literaturnaya Gazeta*, no.
22, May 27, 1970. Condensed text translated and reprinted in *The Current Digest of
The Soviet Press*, 22, no. 22 (June 30, 1970):1–3.

12. That is about 35 percent of the average monthly female salary.

13

PRESCHOOL CHILD CARE
IN ISRAEL

RIVKA BAR-YOSEF-WEISS

The basic model of socialization in industrial societies is a dual system built around two focuses: the family and the formal educational institution. The family, being the procreative unit, is also the first natural socializing agent. When the child reaches a certain age the family ceases to maintain its monopolistic position, and the child enters a nonfamily organization; thus the initial single system changes into a dual system.

Inquiry into educational models in various societies reveals different norms and patterns concerning the "proper" age at which the socialization provided by the family is deemed insufficient and the functions of a second socializing agent are required. In many modern Western societies the first four to six years of the child's life are monopolized by the family. This type of age grouping has been taken so much for granted that sociologists have developed sociopsychological theories showing the functional necessity of a single system, an exclusively family-centered pattern of socialization in the early phases of childhood.[1]

Although the general outlines of the Israeli educational system were patterned after the European model, the Israeli system differs from the European one first and foremost by its tendency to shorten the phase of family monopoly and to begin integrating the child into a dual educational system at an early age. Within this general framework a wide variety of socialization patterns exists due mainly to the range of cultural plurality of the Israeli society and the loose institutionalization of preschool education.[2] The patterns range from the full family monopoly of the preschool period in the rural Arab family to the obligatory dual system starting with the very first days of the child's life in the kibbutz.

The tendency toward the dual system will probably be strengthened in the near future by the extension of the nursery, kindergarten, and day-care types of preschool units and by a growing consensus about the

lowering of the "proper" age limit at which the child is considered ready
to enter a nonfamily unit.

A Short Historical Review

A review of the recent history of the ideas and the institutional development of Israeli preschool child-development programs will reveal both
the major problems that have been solved and those that are still await-
ing solution in the preschool area.

Three periods are distinguishable in the history of preschool education in Israel:

1. The prestate (pre-1948) period when the entire educational system, like
   many other public services, was run on a denominational and voluntary
   basis without state initiative, planning, or control.
2. The first decade of the state when the basic framework of the compulsory
   educational system was built.
3. The second decade and the beginning of the third when the system was
   and continues to be extended, and attention focuses on specific problems
   such as the needs of the very young and of various social groups.

THE PRESTATE PERIOD

Prior to 1948, in the prestate period, the Jewish and Arab communities existed as separate societies each maintaining its own institutions.
This was especially conspicuous in the field of welfare and educational
services. These were generally very limited in the Arab community and,
as far as they existed, were adapted to the norms and expectations of the
traditional Middle-Eastern family and community structure. The Arab
child was cared for and reared mainly by the family. The family was responsible for the child and did not have to account to the community.
There was no real cultural basis for a dual socialization system. Schooling was accepted with some reluctance as a necessary tool of learning
for those who needed or wanted to learn. This meant that schools were
mainly for urban, middle-class male children. The girls stayed home and
their illiteracy was usually considered desirable.

The Jewish community approached education in general and the socialization of young children in particular in a very different way. Formal education of young male children was the accepted pattern of the
most traditional parts of the Jewish community. According to Jewish religious values, ensuring the physical and the cultural continuity of the
Jewish people is a supreme value. If children absorb the cultural heri-
tage and transmit it to their children, they are the natural guarantees of
continuity. The parents and the community are jointly responsible for
the proper education of the children. The father is responsible for the
child's education, but teaching is a specialized job and is done outside

the family. The community provides teachers and facilities and wields the power of informal control. The institutional result of this approach was the widespread custom of sending the three-year-old male child to a special "early school" called the heder. Here a group of children was taught by a male teacher to read the Bible. The heder was a teaching-oriented school, and the child was expected to adapt to the exigencies of often rather primitive physical facilities, discipline, hours of concentration, and sustained learning. For centuries this type of heder was a well-established feature of Jewish communities all over the world. But because of the rigidity of the system, its concentration on religious teaching, and its male orientation, it became obsolete in modern society. The extremely conservative religious sector continued to maintain the heder in its traditional form. The more liberal and the secular sectors of the community rejected the old heder but retained the dual socialization system for the young child.

The secular Jewish community in Israel established its own system of preschool education, which was both a continuation of and a reaction to the traditional pattern. These nursery schools were modeled after the latest educational experiments in Europe, which were strongly influenced by the Montessori method and similar ideas. The schools were against formal teaching and especially against the teaching of reading; instead they emphasized the development of social skills, work and self-help habits, and, above all, a positive orientation toward play. The nursery schools were coeducational and emphasized the principle of equal and similar education for both sexes. But surprisingly, in spite of the tradition of male heder teachers, from the beginning all the nursery school teachers were females.[3]

The nursery schools of Israel's secular community fulfilled some of the functions of the heder in the religious community. The nursery schools, like the heder, were seen as important instruments for the maintenance and transmission of valued cultural contents. In the prestate secular community these were connected with the movement of national revival and consisted mainly of the Hebrew language and the symbols of national identity. In this period of nation building the nursery school provided the basis for a homogeneous new culture that the families with their varied cultural background were unable to do. In the nursery school the children absorbed correct Hebrew and brought home a rich spoken language. Here was developed an original and new childhood culture of nursery rhymes, singing games, dances, jokes, and children's slang. In a well-established culture these are part of the home lore with which the child grows up. They are also part of those cultural memories that create the feeling of common background and identity. The nursery school subculture was not confined within the relevant age group. The children brought its contents home and stimulated their parents to learn with them and to play their part as listeners and active participants. Many of the nursery schools tried to involve the parents by teaching them the child lore. Thus nursery school teachers became an important factor in furthering the development of the new culture. They played a central role in Hebraization, and the seminars for nursery teachers were important cultural centers. The lack of adequate teaching materials stimulated creativity, and many nursery teachers became authors of children's books and writers of songs.

A third factor that enhanced the importance of nursery schools was the concern for the equal status of women. Equality was conceived primarily as economic equality determined by the position of women in the labor market. It was expected that women should have an occupation. Obviously in a society where the extended family did not exist this elicited the need for some solution of the child-care problem. The institutional response to these needs was provided by voluntary women's organizations such as the Organization of Working Mothers, affiliated with the General Federation of Labor, and the Women's Zionist Organization.

The interest of these organizations was centered around the problem of women and children. As part of their routine activities they built and maintained day centers and nursery schools, some of them for special occupational groups such as nurses. In these institutions payment was minimal and adjusted to the economic status of the family.

Of special importance in this period were the health services for mothers and infants. In order to realize the importance of these health services one must remember that this was a hot, underdeveloped Middle Eastern area ridden with all the health risks of this type of society: malnutrition, infectious eye diseases, gastrointestinal infections and malaria, and, as a result, high infant mortality. The Jewish immigrants came to Palestine with a very clear social ideology of building a modern welfare society. One of the main elements of the ideology was the image of the physically strong and healthy "new Jew." Obviously the environment was adverse to this ideal, and creating a new and satisfactory environment was not only a question of providing institutional services but a social challenge depending on the reeducation of the public in general and mothers in particular. This was what prompted the erection of a system of mother- and child-care centers in the early 1920s. The first such clinic was built in 1921 in Jerusalem by Hadassah, a voluntary organization of Jewish women in the United States, and its aim was to teach Jewish and Arab mothers the basics of modern child care and hygiene. The network of mother and child clinics was later enlarged, and many additional centers were opened by Hadassah, the Worker's Sick Fund (Kupat Holim), and various municipalities.

The clinics aimed at preventive care and provided services for the pregnant mother and infants up to their second year. Mothers were taught the proper infant care and the elements of nutrition. The infants were brought to the clinic for a weekly checkup in the first half year and monthly checkups later. The nurses not only demonstrated how to prepare the proper food but often provided the food to be taken home or to be fed to the babies at the clinic.

In spite of the voluntary character of these clinics, they were remarkably successful in introducing modern health practices in the Jewish community. They had a much weaker impact on the Arab community, which did not respond to this type of voluntary approach. No doubt the very spectacular lowering of the rates of infant mortality in the Jewish community was at least partly the result of the activity of the mother-and child-care clinics. Despite the large gap in infant mortality rates between the Jewish and the Arab community, the health standards of the Arab community were also improving, and the infant mortality in the Palestinian Arab community was the lowest in the Middle East.

The proclaimed intention of the mother and child clinics was to provide preventive health services, but obviously these could not be entirely detached from educational practices. Thus the public nurses attached to the clinics were also the agents of indoctrination of educational ideas and of other norms of behavior concerning the home, personal appearance, and proper behavior with infants such as sleeping arrangements, feeding discipline, response to crying, the use or disuse of pacifiers. Perhaps the most important lesson of all was the consequent relationship with a service institution and the readiness to absorb innovations.

Unlike the schools, which were in general sponsored by some public authority, preschool education was open to private enterprise with few or no restrictions. A great number of private nursery schools were run by more or less well-trained kindergarten teachers. These were rather expensive, but they had some advantages. They were small and more intimate. They were within the neighborhood and often allowed for face-to-face relationships with each of the parents.

THE FIRST DECADE OF THE STATE

In the first decade after 1948 there was a marked decline in the importance attributed to preschool education, except for that of five to six year olds, who enjoyed the right to free education stipulated by the law. The period was one of major changes in the institutional framework, the composition of the population, and also in the ideological and cultural underpinnings of the state. In this gigantic readjustment and under the stress of two wars within the decade, attention was focused on universals, problems relevant to the largest groups of the population. Many of the more particular problems, such as the education of the preschool child, were given low priority. New developments occurred only in those areas where some aspect of the preschool system touched upon the two major tasks of this decade: institution-building and the absorption of immigrants.

The transition from a colonial government to an independent state meant the construction of a new institutional structure. Many of the vol-

untary community functions were now taken over by the state. It was
expected that the formerly particularistic services would be replaced by
public ones serving the entire population. A compulsory and free system
of state education was among the important new institutions. The back-
bone of the compulsory education system was the eight-year elementary
school. Preschool education as a whole was not part of the new scheme,
but it was not entirely neglected. Compulsory free education was extended to five and six year olds. For this group special kindergartens
were opened by the municipalities. What was the desirable character of
these kindergartens was much discussed. Whether they should be a
proper "school year" or retain their nursery school attributes was debated. Accordingly there was the problem of location: should kindergartens be attached to schools or maintained as independent units? It was
obvious that physical incorporation into the school premises also meant
"scholastication." For a variety of reasons, not the least among them the
status interests of the professional association of nursery teachers, it was
decided to continue the former tradition and to leave this age group in a
nursery school environment. This also meant administrative autonomy.
The education provided in the kindergarten was mainly a nursery
school type of nonscholastic activity, but there was more emphasis than
before on preparation for school. The existence of these specially designated kindergartens, as a compulsory step toward school, was in itself
fulfilling a preparatory function. The children counted the years and anticipated being enrolled, seeing it as a rite of passage, the first formal
sign of maturation.

The incorporation of one preschool year into the compulsory educational system was an important step forward. It gave full legitimization
to the idea of preschool education and provided at least the beginning
of an organized and controlled system. At the same time, indirectly perhaps, it was instrumental in the decline of the nursery school for the un-
der-five age groups.

In order to ensure the swift application of the law, resources were
reallocated. The voluntary associations and the municipalities cut their
financial involvement in day-care centers and nursery schools for the
younger age groups and transferred their main efforts to the schools and
the compulsory kindergartens. During this period in which the population more than doubled, the publicly sponsored education and care of
younger children did not expand. The absorption of an extremely large
number of immigrants affected nearly every institutional activity. The
difficulty of providing even the most elementary material necessities for
the new immigrants was overwhelming. Food and shelter were obviously the first priorities. But of no less immediate importance were
some other basic problems such as health and hygiene.

The mass immigration endangered the achievements of the prestate
period.[4] A considerable number of the immigrants came from underdeveloped Middle Eastern countries. Many of them had lived in sordid

oriental ghettos under precarious material conditions. They were uneducated and brought with them the attributes of underdevelopment: large families, high infant mortality, low health standards, and lack of hygiene awareness. The living conditions of the transit camps, where the immigrants had to live until proper housing was provided, constituted an additional danger.[5]

The Ministry of Health assumed responsibility for the mother- and child-care services. The existing facilities were extended and new units were added. Mother and child clinics were opened in each transit camp, newly built neighborhood, or settlement.

The number of clinics grew from 120 in 1948 to 658 in 1966, the majority of them (450) opened by the Ministry of Health and a fair number (175) by the Workers' Sick Fund.[6] Special clinics cared for sick and undernourished children.

In all these activities the public health nurses played the central role. They explained and persuaded, cajoled and gave orders, whatever style they considered effective. They also cleaned homes, cooked food, washed and combed children, and distributed clothing. They fought superstition, magic beliefs, and age-old customs that often were detrimental or even dangerous for the infant. Quite often they also fought everything that appeared "different," trying to change the immigrant household as quickly as possible into an average Israeli household. They succeeded quite well in their task of modernizing the immigrants' methods of rearing and caring for children, and undoubtedly they reinforced the new immigrants' aspirations toward conformity.

It seems that the nurses' success can be explained by their enthusiastic belief that they were the representatives of progress and that their task of absorption was of major social importance, and by their lack of doubt as to the correctness of their knowledge. The immigrants reacted appropriately by acknowledging the superiority of the nurse and trusting the magic of modern health practices. The transition did not occur abruptly. Often the old and the new magic were practiced simultaneously. Nevertheless, it may be said in retrospect that the acceptance of the new patterns of behavior was remarkably quick and smooth. The amount of success is most succinctly illustrated by the statistics on infant mortality (Table 26–1). The causes of infant mortality are no less illustrative. In 1950 the major cause of death among Jewish infants was gastroenteritis (7.8 percent); in 1968 death was mainly due to congenital malformations (4.6 percent), while gastroenteritis had nearly disappeared (0.9 percent).

I have tried to show above that the public health nurses were agents of socialization in a larger area than mother and infant care. Their socializing function became much stronger at the time of the mass immigration. The public health nurse was the only official figure who had legitimate rights to intervene in the privacy of the home. Health was used as the lever for changing day-to-day habits, and, given the emotional involvement in maintaining or achieving health and physical well-being,

TABLE 26–1
Infant Mortality Rates
(*per thousand*)

Year	Jews	Non-Jews
1947	29.1	
1949	51.7	67.9
1950	46.2	
1955	32.4	62.5
1960	27.0	48.0
1965	22.7	41.8
1968	20.0	42.4

SOURCE: Central Bureau of Statistics, *Statistical Abstract of Israel,* no. 20 (1969).

NOTE: The incompleteness of the statistics for non-Jews in the early 1950s is mainly due to the negligence of the rural population in registering their infants, especially those who died within the first months. The increase in the percentage of women delivering babies in hospitals and the payment of maternity allowance affected the reliability of these statistics.

it had rare strength. The results were relatively immediate and extremely satisfactory. The sight of well-fed, clean, and healthy babies was the best reinforcement for changed behavior. The nurses in their zeal as agents of socialization tried to introduce total modernization. They taught not only nutrition and cleanliness, but also how to dress, what to buy, how to arrange the home. They also tried to convey their own ideas on the rights of women against the tyranny of patriarchal, traditional husbands.

As was to be expected, the modernizing tendencies were more easily accepted in those cases where the new pattern was linked to routine behavior and less successful when they touched deep-rooted beliefs or interaction systems. One critical example of the effects of partial modernization was the size of the family. Neither the nurses nor the institutions connected with health and welfare had a clear idea about the desired policy of family planning. The typical Israeli family was a modern, small nuclear family. There were two exceptions—the Arab and the very religious Jewish family. These constituted a variant pattern of large, traditional families. Paradoxically the official ideology was for large families, partly as a political gesture toward the religious values and partly motivated by the feeling that the extremely high rate of reproduction of the Arab population had to be counterbalanced. This ideology did not affect the modern parts of the population, but it legitimized in Israeli terms the values and customs of the traditional groups, notably the immigrants coming from the underdeveloped Islamic countries.

The public health nurses were in a dilemma. They believed in planned families and considered the constant pregnancies of these new immigrant mothers as a sign of backwardness, but they felt some responsibility toward the official policy, which was against family planning. Some of them tried to explain the importance of planning, but they were rebuked by the official authorities and were met with hostility by the immigrants, especially the husbands. At the same time the institutional structure was not geared to the absorption of very large families, and the immigrants themselves were unable to manage their families under the conditions imposed on them by modern urban life. The majority of the families adapted themselves to the behavioral norms observed by the old-timers and the Western immigrants and chose to plan their families. But about 25 percent of the families maintained the old pattern of unplanned families. Because of the high rates of survival of the infants, these families were much larger than the nonplanned families in the immigrants' countries of origin. The outcome was predictable. Nevertheless, it was unexpected, and Israeli society was not prepared for it. Many of these large families slid into the vicious circle of poverty: large families, low education of the father, low per capita income, low level of scholastic success of the children, poor occupational opportunities, and hence continuity of poverty.

In the 1950s and early 1960s there was much reluctance to discuss openly this profile of poverty. There was a pervading belief that since the parents' generation could hardly be changed, it had to be helped as a temporary measure; the situation would disappear in the next generation. Compulsory education was seen as the universal magic solution. It was believed that education would provide the basis for equal chances. The compulsory kindergarten year was seen as an extra effort to ensure the adequate preparation of all children for school. It took more than a decade to realize the inadequacy of the universal educational system in eliminating the social and cultural gaps. Unfortunately, by neglecting the expansion of prekindergarten nursery schools for three to five year olds, the very instrument that could best serve as preparation for school was relinquished.

The complex problem of differences in learning abilities of various cultural groups was nearly taboo. It touched the very sensitive area of the image of the Jewish people and the success of the Israeli social structure. It was stated that "the Jews are the same wherever they lived in Diaspora," and "absorption is only a question of time." It was a major disappointment to all concerned that these beliefs were proven to be unjustified.

The problem of differential intellectual development had been recognized by various scholars in the first years of the mass immigration. Research focused on determining the exact quality and quantity of these differences, the age at which they become discernible, and their genesis.[7] Remedial programs were devised and suggested.[8] The research findings converged into two important points:

1. The differences in intellectual development are well discernible at the prekindergarten age; they are measurable at the age of two or three and probably even earlier.
2. Linguistic competence is a very important element of the differential development, and it may be the central factor, for the low level of linguistic competence of the children of immigrant families coming from underdeveloped countries weakens the ability of these children to profit from kindergarten and school education.

FROM 1960 TO 1970

During the last decade there was a growing awareness of the complexity of intercultural differences. It was also realized that the universalistic approach of equal treatment through formal institutions constitutes de facto discrimination, harming the children's chances for equal integration into all areas of the social structure. One of the most obvious indices of this situation was found in education. The last decade was characterized by the endeavor to introduce differential approaches toward socioeconomic and cultural groups in the fields of welfare and education. There is a shifting of emphasis from equal treatment to equal outcomes. From the institutional viewpoint fewer difficulties hinder implementing a policy of differential treatment than during the period of mass immigration. The institutional structure and the basic universalistic norms of its functioning are now stabilized; therefore, there is more susceptibility to special cases and more elasticity in allowing the simultaneous implementation of heterogeneous programs.

The readiness for program variants necessitated the reevaluation of the importance of various social problems and decisions on new priorities in the allocation of resources. Within the framework of this reassessment, attention turned once more toward the care and education of the preschool child. Three different high priority needs seemed to converge in the problem of the preschool child.

1. A relatively high percentage of children whose parents immigrated from Arab countries were failing scholastically. Since the last great wave of immigration occurred in 1956, these children who were born and socialized in Israel exposed a weakness of the system of absorption and integration. Research and plain common-sense observation persuaded the authorities that the prekindergarten age is crucial in preparing the child for school.

2. How could the problem of poverty among large families be dealt with? The general rise in the standard of living accentuated socioeconomic differences and revealed the inability of certain groups to keep pace with the general trend of increased material well-being. The fact

that the majority of the poor were concentrated in a well-defined socio-cultural group only aggravated the general feeling about the situation. In general the profile of the poor could be traced as slum dweller, uneducated, with a large family, of North African origin, having one wage earner in the family who was unemployed, occasionally employed or employed in marginal, very low-income occupations. Obviously the strength of this syndrome lays in the high interdependence of its variables, and it is of no importance that many of the slums are newly created by the families themselves. Among the various programs aimed at attacking the poverty syndrome, some were directly relevant to the care of the young child. It was assumed that the most efficient way to assure the physical and mental well-being of the slum children was to keep them in well-provided day-care centers during the day. This would also relieve the harassed mother from caring for many small children and eventually make it possible to draw her into the labor market, thus considerably raising the family income.

3. In the present situation of scarce labor, women constitute a major potential labor resource. The low percentages of working women show that these resources are only superficially utilized.[9] Lately attempts were made to draw women into the labor market and to prevent their leaving after marriage. But until there is a relatively inexpensive solution for the care of the children of working mothers, these attempts appear somewhat futile. The Ministry of Labor, in answer to this problem, decided to open and support day-care centers for the children of working mothers. Although the majority of the working mothers are currently neither from the low-income groups nor mothers of large families, it is hoped that the extension of the net of day-care centers for working mothers will motivate women from these groups to seek employment.

The institutional responsibility for the well-being of the preschool child involves several ministries, public institutions, and voluntary organizations.

The Ministry of Labor acts in a double role through the Institute of Social Security and through the Department for Working Women. The Institute of Social Security is mainly responsible for the implementation of the legal requirements for the insurance of the pregnant mother and the infant and the payment of maternity and family allowances. The newly established Department for Working Women is involved in the problem of day-care centers for working mothers. The Ministry of Labor assumed the administrative and financial responsibility for opening new centers and enlarging some of the existing ones. In certain cases it pays a special allowance to mothers who pay for the care of their preschool or prekindergarten children.

The Ministry of Education is responsible for educational control over kindergartens and nursery schools and for training kindergarten and nursery teachers. The Ministry of Health is maintaining many mother and child health care centers. It also provides inoculations and is responsible for the health standards of the nursery and kindergarten schools. The Ministry of Welfare takes care of the special cases of the chronically ill, the retarded, the mentally ill, the orphans, and the abandoned children.

The Office of the Prime Minister is a new arrival in the field. In 1968 it was decided to establish the Demographic Center as part of the prime minister's office. The task of the center was to coordinate the activities pertaining to the care of the preschool child and family planning, to design long-range plans, and to suggest improvements in the existing system. The center is helped by a public council and an executive committee. It does not have executive authority, and its work is limited to the initiation and coordination of "interministerial and interagency consultation on a regular basis . . . concerned with strengthening . . . support of the family to ensure priority consideration for all those services which help the child to become a self-sufficient and well-integrated, constructive member of an expanding society." [10]

The municipalities repeat the governmental pattern, and their various departments engage in the same types of activities. In general the implementation of governmental projects is carried out by the corresponding municipal departments. But this is not necessarily the case. There are certain activities for which state governmental departments are directly responsible.

Voluntary organizations play an important role in the area of child care. Most important are the activities of the organizations affiliated with the General Federation of Labor (Histadruth):

1. The Worker's Sick Fund, besides being the major provider of health insurance and health services, is also responsible for the preventive and curative care of pregnant mothers, hospitalization at delivery, and clinical health services for the children of about 80 percent of the population. It also maintains many mother- and child-care clinics.
2. The Organization of the Working Mothers runs many of the day-care centers and prekindergarten nurseries.
3. The Histadruth with the Collective Settlements (these are organizationally part of the Histadruth) maintain several colleges for the training of kindergarten and nursery teachers.
4. The Professional Association of the preschool teachers sets the training standards and controls the professional activities of its members. It is endeavoring to develop a system of knowledge, educational contents and values pertaining to the preschool child.

Organizations such as the Women's International Zionist Organization (WIZO) and the Organization of Religious Women are also supporting and managing preschool units and teachers' training institutions.

While this proliferation of authorities may augment the resources of those involved in child care, it is also to be blamed for the lack of coordination, the absence of well-designed overall planning, and the many inconsistencies of the system. The Demographic Center was established to correct the faults inherent in such a loose pluralistic system. But by its structure the center is incapacitated and unable to fulfill this function. Being devoid of real executive authority and having a very limited budget, it must restrict its activities to brainwork and persuasion. As such it does important work. It is the first time that a central, qualified body took a global approach to the subject, saw the system as a whole, and prepared plans accordingly. While this is a necessary preliminary phase in tightening the system, it cannot be considered sufficient.

Current Care for the Preschool Child

Let us now survey the current Israeli pattern of care for the preschool child.

THE PREGNANT MOTHER

The prenatal care of the child implies care for the health of the mother, provision of economic and service resources, and training for the maternal role. The health care is given by the clinics of the Worker's Sick Fund and other smaller medical insurance schemes, which cover about 90 percent of the total population. Specialized preventive care is provided by the mother and child clinics. On the average about 60 to 65 percent of pregnant women are under the supervision of these clinics.[11] The great majority of deliveries occur in hospitals. The health authorities consider this as a major success in ensuring the health of the mother and the infant.

In order to increase further the rates of hospital delivery, maternity allowances are paid only to hospitalized women. At present 99.9 percent of the Jewish women and nearly 90 percent of the Arab women, including the nomadic Bedouins, have their children in hospitals. It is worth emphasizing that ten years ago only about 50 percent of the Arab women went to hospitals. Undoubtedly the low level of infant mortality is correlated with the practice of hospital delivery.

The economic aid to the pregnant mother includes free hospitalization, a maternity allowance, a birth payment, and the right to return to her work place after three months or one year. The first three of these schemes are included in the general and compulsory social insurance. Insurance fees are paid by all earners. Hospitalization is by definition part of the insurance, and the hospitals are paid by the Institute of Social Security. But de facto hospital delivery became a service as failure to pay the insurance rates does not forfeit one's right to it. The maternity allowance is a lump sum intended to provide for the purchase of the layette and basic necessities for the infant. The birth payment is paid to working mothers who have the right to three months' paid absence from work. Additional laws ensure the rights of a working mother to an unpaid absence from work of up to one year.

The mother and child clinics also have a program to prepare parents to care for their children. There are formal courses for parents-to-be, but fathers do not usually attend these courses. The important part of the socialization is informal, as a result of contact with the nurses and with other pregnant women. The men do not take part in these situations. Thus in spite of some attempt to prepare women and men for their parental roles, it is the women who go through the process of socialization partly forced upon them by the physical fact of pregnancy, while the fathers stay outside the process that for them starts, at best, with the birth of the child.

THE INFANT

The burden of guiding the mother and providing preventive care of the infants is assumed by the mother and child clinics. In 1967 about 85 percent of the infants and about 70 percent of the children between the ages of one to four were in more or less continuous contact with the clinics. In 1968 there were 679 clinics, about 459 maintained by the Ministry of Health, 185 by the Workers' Sick Fund, 30 by Municipalities, and the rest by voluntary organizations.[12] The free services of the clinics are guaranteed by the Ministry of Health, but use of the services is voluntary. The high percentage of users can be explained by the force of informal persuasion, the work of the public health nurses, the positive image of the clinics, and the tendency toward conformity. The clinics are a part of the "pregnancy culture" of Israel.

The following detailed description of the typical pattern of activities of a clinic in the case of a first pregnancy will help to illustrate the nature of its work:

As soon as the young woman becomes aware of her pregnancy, she is usually encouraged by either her friends or her physician to attend the maternal and child-care center in her neighborhood. (Though the centers try to get women to come during the first months of pregnancy, the majority nevertheless still wait until the fifth or sixth month as there is not yet enough awareness of the importance of the first trimester.) In addition to undergoing a physician's examination, a medical and social history, and blood tests (Wassermann, Rh), she will then receive instructions from the nurse regarding proper nutrition, and in areas of economic deprivation the clinic may also supply some nutritional supplements if necessary. Most usually the expectant mother will then join a group of other young women, where under the guidance of the nurse she can discuss all aspects of pregnancy and childbirth that concern her. These discussions may range from specific ethnic beliefs and "grandmother's tales" to individual phobias and fears, to actual questions regarding the biological and medical aspects of pregnancy and birth. Thus these discussions help lower anxiety and also help attain a greater degree of continuity between the beliefs, expectations, and practices of the family and the professionals. There is a tendency to encourage the expectant mother to breast-feed her child—both for emotional and for medical reasons—and there is also a trend to make available courses and exercises in natural childbirth. (The use of anesthesia in childbirth is minimal and usually limited to pathological deliveries.) If there are any medical problems during pregnancy or special risk conditions, they will be recorded and the hospital will be alerted by the clinic. (A complete medical record is kept of each mother and child, beginning with the first visit to the clinic. However, the records are handwritten, not standardized, coded, or punched, and are in need of revision in order to be useful for efficient data collection or centralized registration.)

As part of her preparation for delivery, the expectant mother will often be taught the bases of child care as well—how to diaper, how to nurse, and so forth. Routine visits are monthly through the seventh month, twice monthly in the eighth month, and weekly in the ninth month. Thus the prenatal care encompasses medical care, nutritional supervision, education, and emotional preparation.

As soon as the young mother returns from the hospital with her baby, "her" nurse will make a home visit. (A close relationship usually develops between the family and the nurse, and not infrequently the nurse might be the same person who cared for the expectant mother when she was a child herself. In many instances this long-term intimate relationship is the source of the nurses' special ability to help the family, particularly in the lower socioeconomic groups where her role becomes central in the life of the family.) At the first home visit the nurse will check the conditions in the home, help the mother with problems of nursing and diapering, and invite her to the clinic for the initial checkup on the fifth to eighth day after birth. (If the mother fails to turn up, she will be visited at home again.) The first checkup is a general one made by the nurse, and in some centers it now includes a hearing test (which is also repeated at five and seven months.) Afterward the child is brought again for an examination by the pediatrician, who then continues to see the child routinely once every three months. In some clinics there is now a growing tendency to have the nurses also do some rudimentary developmental screening of sensorimotor and social items at approximately three, six, twelve, and eighteen months—but this again is not yet standardized or routine in all clinics and is usually dependent on the interests of the pediatrician in any particular clinic.

Inoculations are initiated at three months and include vaccines against smallpox, diphtheria, whooping cough, tetanus, poliomyelitis, and German measles. The routine visits to the clinic are planned around the terms of inoculations—usually three, six, nine, twelve, eighteen months, and two and three years.

When coming for inoculations, the mother will also have a short talk with the nurse, possibly about herself, the child, or even other children in the family. (A multiparous mother may bring several children at the same time.) This is an opportunity for the nurse to pick up behavioral and emotional problems, either through the communication of the mother or through observation of the actual behavior of the mother and the child. If such problems do exist, the better trained nurses will try to discuss them with the mother, and in those instances where there is a consultant psychologist in the clinic (in a very small minority of the clinics), she will refer the problem to the psychologist or consult the psychologist herself and receive guidance on how to help the mother. In some clinics the mother may be put into a discussion group led by the nurse.

If there is a more complex medical problem, such as a multiple handicap in the child or postpartum depression in the mother, the nurse may attempt to obtain the necessary curative services (though these have to be given by other agencies).

A recent innovation is a program to encourage mothers to "speak" to their infants. The health nurse explains the importance of verbal communication in early speech development, and the mothers are given a booklet called *Speak to Your Child* (G. Ortar).

During the second year of life the contact with the clinic diminishes, and in the third year it is usually minimal. This is considered a weak point in the present program resulting from a failure to educate the public about the necessity of continued physical and psychological supervision during the preschool years. Some clinics are now thinking of how they can involve themselves in the neighborhood day-care centers and kindergartens, but this is still a question of the future.

When the child goes to school, his records are turned over to the school health service. In some instances the same public health nurse serves both in the child-care clinic and the school, and in such a case there is a certain continuity through which she can truly follow the child during his complete course of development.[13]

Statistics and qualitative evaluations show that the clinics are well established, and that in spite of their voluntary aspect the use of their services is nearly universal. There is also general consensus about the success of the clinics in inculcating standards of hygiene, modern feeding methods, and awareness of the physical development of the children. Their weakness lies in the nonphysical aspects of the child care, which could be helped by the clinics, but neither the staff nor the routine activity are geared to it. "Only minimal progress," Joseph Marcus notes, "has been made towards reorienting the professionals of these centers towards a style of work which focuses on the mental health needs of children, regular developmental diagnosis and the early detection of deviation. Also, little has been done, as yet, to provide the staff with sufficient training in developmental medicine, and nothing has been done to standardize any developmental test so that there would exist normative data for Israeli children."[14]

The high prestige of the clinics and the number of users makes them the natural instruments through which to overcome problems of backwardness that are caused by lack of early stimulation. This seems a much more difficult task than health care, as there are no established standards and routines that can be transmitted in a nearly mechanical manner. But it does not seem impossible. Standards and simple routines can be developed, stimulative material can be distributed, parents can be taught some elementary rules, and the control or stimulation of mental development can be institutionalized in the same manner as the control of physical development.

From the mother's viewpoint the great problem is the lack of facilities for the care of the infants of working mothers. There are only a few day-care centers that are ready and equipped to receive infants. At present there are day-care facilities for approximately 1 percent of the infants of urban families. Mothers are often reluctant to take their babies out of their homes. The lack of facilities, the popularity of the psychoanalytical emphasis on the mother-child relationship, the Spock ideology, and the home-centeredness of the traditional family had a mutually reinforcing influence in creating the norm that "a mother should care for the infant at home." Working mothers often prefer to stop work for the period of family building, to stay home for the first year after birth on unpaid maternity leave, or to employ a paid household helper who besides caring for the infant also does some housework. For these reasons it is estimated that the demand for day-care facilities for infants does not amount to more than an additional percent—a total of 2 percent of infants under one year.[15]

The strength of the desire to keep the infant at home is manifest even in the kibbutzim where lately there is a growing tendency to move the children from the separate children's houses into the homes of their parents.[16] In the urban areas, on the other hand, there seems to be a slow weakening of this reluctance and more readiness than before to accept a dual system of socialization at an early age.

The extrafamily institutions now train and help the family, but the family remains, in the majority of cases, the exclusive agent for the care of the infant. The institutional approach to the care of infants is focused on the physical well-being of the child. In spite of some new developments, health, hygiene, physical growth, and habits remain the principal functions of this part of the formal system.

THE YOUNG CHILD

For the next chronological period of the child's life there does not exist an institutional structure parallel to that of the mother and child clinics. The importance of the clinics decreases after the first year, although they continue to accept responsibility for the preventive aspects of health care. But the attendance of the clinics decreases, for the general habits of hygiene are by that time thought to be established and the child becomes more resilient. The curative problems are dealt with by the pediatricians of the various sick funds. With the weakening of the ties between the mother and the clinic, the dual system shrinks. There is no constitutional continuity of the extrafamily system, no redefinition of developmental functions, and no comparable cultural norms for the use of services. From this viewpoint there is a hiatus in the structure of the system, which is then reclosed for the five to six year olds with the compulsory kindergarten. As shown above, the one to five age period is not a total institutional void, but its function, structure, necessary conditions, and resources are unsystematic, occasional, and badly defined. Preschool education is not a continuous process built in chronological steps from birth to the formal school system. Rather preschool education historically began at the two ends of the period—the infant-care stage with its physical definition and the kindergarten stage with its educational, quasi-scholastic approach. Recent developments favor a stepwise extension of the system chronologically downward. After the compulsory kindergarten a prekindergarten network is to be built for four to five year olds and three to four year olds.

Besides the differences in the definition of their functions there is a major methodological difference between the kindergarten and the mother and child clinics. Because the primary care of infants is generally considered to be in the hands of the mothers, the mother and child clinics serve an auxiliary function, training the mothers to carry out their tasks. The kindergarten educates the child. Its function is not auxiliary; it is complementary. Hence it pays little attention to the role of the parents and their ability to fulfill their part in the dual system. The complementary activities depend very much on the assumption that the home brings up the child to a certain standard. Without this standard the complementary education loses much of its impact and meaning. In order to prevent school failure, there is a tendency to transfer to formal education some of the educational functions of the home and thus to serve as a substitute for the unsatisfactory informal environment. According to this tendency there are two factors that shape the content of the preschool education:

1. The complementary element, whose function is the enlargement of home education by adding dimensions that are by definition extrafamiliar
2. The element of substitution, which should remedy the lacuna left by underprivileged homes, assuming that the middle-class educational standards serve as a reference point [17]

The complementary aspect is well developed as it was the principal topic of preschool education prior to the 1960s. It is also the main target of the training programs for preschool teachers. The substitution aspect is a recent addition, more problematical than the former, but it is now given high priority because of its relevance to social integration and equality. The kindergartens, like the schools, encompass nearly 100 percent of the relevant child population. In the majority the day is four or five hours, and long day kindergartens are the exception. All the teachers are females. Their median age is thirty; about 50 percent are Israeli born; and more than 70 percent are qualified by some recognized, two-year kindergarten teacher's college.

About 50 percent of the prekindergarten age group (three to five) attend schools. About 80 percent of these are public nursery schools, and the remaining 20 percent are private schools. The public schools cater first of all to the underprivileged. Thus in the immigrant villages 100 percent of the children attend prekindergarten schools. In the development towns 80 percent attend. The well-developed urban areas are in a much worse position.[18] The cities were given low priority in the establishment of tuition-free prekindergarten classes. The poorest neighborhoods were the first eligible for the program. The lower-income groups slightly above the poverty line were adversely affected by this policy. They are unable to bear the cost of maintenance, which is required even in public institutions from those who are ineligible for the "poor status." This situation might now change. In 1971 the Minister of Education announced the enlargement of the program to provide places for *all* children three to five years old. There will be a sliding scale payment based on the income of the parents and the size of the family.

One to three year olds are most neglected. Two hundred and eighty day nurseries, about 90 percent of which are long-day nurseries with meals, cater to approximately 10 percent of this age group.[19]

There are no reliable statistics about the staff of the prekindergarten schools. According to some impressionistic observations, the training of the child workers is varied and haphazard.[20] Usually each voluntary organization maintaining a prekindergarten program trains its own workers. Recently the Demographic Center started to develop a standard program for training prekindergarten teacher-nurses. The program is a combination of health, child-care, and educational skills. Much attention is given in the program to music and handicrafts. Interestingly one of the smallest items (ten hours) is the guidance of parents.

As far as forecasting is possible, in the next ten years the school attendance of prekindergarten children, mainly three to five year olds, will be increased, and there is a high probability that parallel intrafamily and extrafamily education will become the general pattern. It is also expected that many of the kindergarten and prekindergarten schools will be open for a long day (nine to twelve hours). The number of day-care centers for children under three will increase, mainly for children of working mothers and possibly in connection with work places or special occupations.

The activity programs are undergoing standardization; the applied research that has already started will be expanded and concentrated mainly on intellectual and social stimulation programs. Training is becoming professionalized. The topic of preschool education is moving into the academic sphere. The Hebrew University has already started a small-scale program of teaching and research. This will undoubtedly grow into more systematic interest and investigation about the needs of this age. The educational and administrative control of the central administration is slowly being extended, and it seems a safe prediction that it will tend toward at least supervisory authority over the system as a whole.

INSURANCE AND ALLOWANCES

The developing interest of the state in the welfare of the children strengthened tendencies to provide economic help to families. The income maintenance approach of the child-welfare policy represents a revival of ideas that were the accepted pattern in the prestate period.

Before the accelerated economic and industrial development and the mass immigration, the wage system was based on "family wages" determined by the number of dependents in the family. This wage system created a handicap for the head of large families. Employers preferred the cheaper worker with a small family. With increased industrialization occupational differentiation gained in importance, and the family wage system became obsolete. But the problem of the standard of living of large families was even more acute than in the preindustrial phase. The new solutions are based on the principle of differentiation between the producer role, and the wages and benefits linked with the producer's position in the occupational system, and the consumer role, seen as the expression of the welfare position. Egalitarian tendencies are found in both areas and are expressed in a policy that tries to reduce income differentials and to create a welfare policy that secures a minimal basis of consumption for each member of the society, independently from achievements in the labor market. Based on this conception we are now in the phase of the return of family allowances, but this time they are paid by the Institute of Social Security. At present there are two separate schemes, both compulsory:

1. The workers' children insurance is a redefinition of the former family wage. It covers the first three children of all employed persons. The insurance fees of 1.8 percent of wages and salaries are paid by the employer.
2. An allowance for large families is paid for the fourth and following children of all earners. In the case of the employed the employer pays the insurance as an additional 1 percent of wages and salaries. The self-employed pay 1.5 percent of their income as insurance.

The workers' insurance is basically a part of the earner's fringe benefits, but in such form that it eliminates the danger of discriminating against heads of families. The allowance for large families is a more general scheme, and it has no link with occupational activity.

The policy of family allowances suffers from an ideological dilemma caused by the lack of clarity about the desired aim of family planning. If one assumes that the allowance has some influence on the number of children a family plans, then it is of major importance to decide which is the desired norm. Obviously the workers' children insurance envisages a three-child family, while the large family allowance is a sort of incentive for having more than three children. The integrative and egalitarian values, if taken seriously, imply more uniformity in the family size of various socioeconomic groups. No welfare policy will be able to eliminate the problems created by very large low-income families. At the same time there are strong values favoring large families. Obviously if there is a clarification of the official attitude and if the accepted ideal is to be a three- or four-child family, it will be necessary to maintain two different policy schemes, one for existing families that should not be affected by this kind of policy and a second scheme that should eliminate incentives for having very large families. The probability of such a policy is not high, and it seems that the official policy will continue to be ambiguous and nondifferentiated. But now for the first time there is open discussion of the relationship between child welfare and family planning. There is also some awareness among the groups with traditional cultures that the readiness for family planning might be the key to the self-help of the underprivileged stratum.

Conclusion

The central question of preschool care seems to be the integration of system and policy. It seems reasonable that to achieve such coordination a central authority should be developed at the ministerial level. Such an authority does not necessarily have to assume monopolistic rights in the field. But it would need to develop long-range and short-range planning, to have necessary resources, and to exert supervisory authority. It does not appear that the Demographic Center will develop into such a Children's Office. It is rather difficult at present to envisage the situation in which the particularistic interests of various ministries, departments, and organizations will cease to circumvent coordination and integration. But in order to conclude on an optimistic note I shall repeat the much-used Israeli formula, "Israel is a land of miracles where anything can happen. Things will turn out right."

NOTES

1. Talcott Parsons, "Family Structure and the Socialization of the Child," in T. Parsons and F. Bales, *Family Socialization and Interaction Process* (New York: The Free Press, 1955).
2. I shall use the term "preschool" to designate the under-six age group. This, according to the official definition, is the age at which the Israeli child is allowed and required to enter the first grade.
3. The kibbutzim experimented in this early period with the idea of placing men in the nurseries, but they did not succeed and abandoned the idea. Men do work in the nurseries occasionally as replacements on holy days.
4. S. N. Eisenstadt, *The Absorption of Immigrants* (London: Routledge & Kegan, 1957), and S. N. Eisenstadt, *Israeli Society* (New York: Basic Books, 1967).
5. During 1950 to 1952 there were 250,000 immigrants in transit camps. Statistics of the Jewish Agency, Department of Immigrants.
6. J. Taustein, "Mother and Child Health," *Health Service in Israel* (Jerusalem: Ministry of Health, 1968), pp. 116–150.
7. Dinah Feitelson, *The Causes of Failure in the First Grade* (Jerusalem: Henrietta Szold Institute, 1953); Gina Ortar, "Comparative Analysis of the Structure of Intelligence in Various Ethnic Groups" in Carl Frankenstein, ed., *Between Past and Future* (Jerusalem: Henrietta Szold Institute, 1953).
8. Lea Adar, "Learning Difficulties of Immigrant Children," *Megamoth* 7, no. 2 (1956): 1939–80 (in Hebrew); Gina Ortar and Carl Frankenstein, "How to Develop Abstract Thinking in Immigrant Children from Oriental Countries," in Frankenstein, *op. cit.*, pp. 291–316.
9. About 30 percent of the Jewish women are in the labor force, but only 25 percent of the married women are working. The percentage of Arab working women is very low.
10. Zena Harman, *Services for the Young Child* (Jerusalem: Demographic Center, Prime Minister's Office, 1969).
11. Taustein, *op. cit.*
12. *Statistical Abstract of Israel*, no. 20 (Central Bureau of Statistics, 1969), pp. 597.

13. Joseph Marcus, "Early Child Care Programs in Israel," unpublished manuscript (Jerusalem: Henrietta Szold Institute, 1970).

14. *Ibid.*

15. Harman, *op. cit.*

16. An analysis of the meaning of duality in the kibbutz education is found in Rivkah Bar-Yosef, "The Pattern of Early Socialization in the Collective Settlements in Israel," *Human Relations* 12, no. 49 (1959): 345–360; cf. Melford E. Spiro, *Children of the Kibbutz: A Study in Child Training and Personality* (New York: Schocken Books, 1958); Leslie Rabkin and Karen Rabkin, "Children of the Kibbutz," *Psychology Today* 3, no. 4 (September 1969); Joseph Shepher, "The Child and the Parent-Child Relationship in Kibbutz Communities in Israel," *Assignment Children*, no. 10 (United Nations Children's Fund) (June 1969), pp. 47–56. Publications Department, Ichud Habonim, *Kibbutz: A New Society? An Anthology* (Tel Aviv: Ichud Habonim, P.O.B. 3214, 1971).

17. Sarah Smilansky and Moshe Smilansky, "The Role and Program of a Kindergarten for Socially Disadvantaged Children" (Jerusalem: Henrietta Szold Institute, 1969), publ. 490.

18. *Statistical Abstract of Israel.*

19. Harman, *op. cit.*

20. Joseph Marcus, *Day Care for Infants and Children* (Jerusalem: Henrietta Szold Institute, 1970).

Appendix

JOURNALS AND ORGANIZATIONS*

Journals

These publications are listed primarily as a source of articles, reviews, studies, and comments and are available in libraries, especially those connected with universities.

Child Care Quarterly. An independent journal of day and residential child-care practice. Published by Behavioral Publications, Inc., 252 Broadway, New York, N.Y. 10025. Annual subscription $12.00 for individuals, $25.00 for institutions.

Child Development. Published by the Society for Research in Child Development, University of Chicago Press, 5750 Ellis Ave., Chicago, Ill. 60637. $15.00 per year. One of three publications of the Society. The others are monographs of the Society for Research in Child Development ($12.00 per year) and Child Development Abstracts and Bibliography ($8.00 per year).

Childhood Education. Published by the Association for Childhood Education International, 3615 Wisconsin Ave. N.W., Washington, D.C. 20016. Membership in the organization ($6.00 regular, $2.50 student) includes the magazine.

While focused on schools, it generally has a point of view of value to all professions working with children.

Children. Issued six times a year by the Children's Bureau, U.S. Department of Health, Education and Welfare. To subscribe send $1.25 to Superintendent of Documents, Government Printing Office, Washington, D.C. 20402.

An interdisciplinary journal that carries articles on topics of interest to those who work with children and families and also gives news of current developments, new books, and pamphlets.

Child Welfare. Journal of the Child Welfare League of America, Inc., 44 East 23rd St., New York, N.Y. 10010.

A professional journal concerned with the welfare of children, with practical methods, research, and education as they relate to child-welfare services,

---

* List and annotations assembled in part by the Office of Child Development, U.S. Department of Health, Education, and Welfare.

and with issues of social policy that have a bearing on them. Published ten times a year. Subscription $5.00. Single issues $.75.

Day Care and Child Development Reports, 2814 Pennsylvania Ave., Washington, D.C. 20007. A bi-weekly newsletter on child-care legislation and happenings.

Exceptional Children. Published by the Council for Exceptional Children, 1201 16th St. N.W., Washington, D.C. Published ten times a year. Available to members at $8.50 per year; to agencies and libraries at $10.00. Also issued is a quarterly, *Education and Training of the Mentally Retarded.* $5.00 a year. Both are for professionals.

Report on Preschool Education. Bi-weekly news service about important events in the area of early childhood development. Published by Capitol Publications, Suite G-12, 2430 Pennsylvania Ave., Washington, D.C. 20037. $40.00 per year.

Resources for Child Care. Bulletin containing listings of manuals, articles, books and posters concerning child care. Published by Day Care and Child Development Council of America, 1401 K St. N.W., Washington, D.C. 20005.

Review of Child Development Research. Published by the Russell Sage Foundation, 230 Park Ave., New York, N.Y. 10017. Published irregularly; price $8.00 per year.

Young Children. Issued six times a year by the National Association for the Education of Young Children, 1834 Connecticut Ave. N.W., Washington, D.C. 20009. Members of NAEYC receive *Young Children* as part of their membership privileges; nonmembers may subscribe for $5.00 annually.

Articles of interest to teachers and others working with young children (preschool and nursery) and their parents in day-care centers, camps, nursery schools and other settings.

Other Sources of Information, Film Lists, Bibliographies

American Academy of Pediatrics, 1801 Hinman St., Evanston, Ill. 60204.

American Rehabilitation Foundation, Institute for Interdisciplinary Studies. "Day Care Reference Sources—an Annotated Bibliography." 1800 Chicago Ave., Minneapolis, Minn. 55404. 1970, 29 pp., $2.00.

American Social Health Association. "About Family Life Education." 1790 Broadway, New York, N.Y. 10019. Single copies free. Compilation of resources available to those planning educational programs. Topics include national resources, film resources, study centers, curriculum guides, and journals.

Association for Childhood Education International, 3615 Wisconsin Ave. N.W., Washington, D.C. 20016. Publications and other information for those concerned with children two to twelve.

Bank Street College of Education, 610 W. 112 Street, New York, N.Y. 10025. Publishes a listing of packets, books, booklets, and articles for teachers and parents, available through their bookstore.

Black Child Development Institute, 1028 Connecticut Ave. N.W., Suite 514, Washington, D.C. 20036. Evelyn K. Moore, Director.

Child Development Associate Consortium, Inc., 3615 Wisconsin Ave. N.W., Washington, D.C. 20016.

Child Study Association of America, 9 East 89th St., New York, N.Y. 10028. Publishes a listing of selected books and pamphlets for parents and professionals, some of which are produced by the CSAA staff and others are selected by them for their value and utility.

Child Welfare League of America, 44 East 23rd St., New York, N.Y. 10010. Publications list. Contains books and pamphlets reprinted from journals or published by the League on such topics as administration, adoption service, child development, day care, foster care, group care, homemaker service, and services for unmarried parents. The CWLA has also published standards for services in seven areas such as adoption, foster family care, child protective service. Teaching aids, record forms and film lists are also available.

The Children's Lobby, 112 East 19th St., New York, N.Y. 10003. National lobby for children founded in 1970.

Day Care and Child Development Council of America, Inc., 1426 H St. N.W., Washington, D.C. 20005. List of publications available upon request.

The Education Commission of the States, Task Force on Early Childhood Education, 1860 Lincoln St., Denver, Colo. 80203.

Education Development Center, 55 Chapel St., Newton, Mass. 02158. EDC has a large number of free pamphlets available on day-care and early childhood matters.

Educational Research Information Center (ERIC), University of Illinois, 805 West Pennsylvania Ave., Urbana, Ill. 61803. ERIC is a nationwide service for teachers, administrators, and researchers who are seeking results of recent educational research and demonstration projects. Pertinent information on these areas is announced in *Research and Education,* a monthly publication. For subscription write to Superintendent of Documents, Government Printing Office, Washington, D.C. 20402. $11.00 annually, or $1.00 per issue.

National Association for the Education of Young Children, 1834 Connecticut Ave. N.W., Washington, D.C. 20009. List of publications free from NAEYC.

Office of Child Development, Project Head Start, Post Office Box 112, Washington, D.C. 20013. "Films Suitable For Head Start Programs."

Office of Economic Opportunity. "OEO Film Guide," 1970. Publications Distribution, 5458 Third St. N.W., Washington, D.C. 20011. This listing of more than 150 films includes films produced by the Office of Economic Opportunity, TV documentaries, and a variety of poverty and poverty-related classics. It includes a large section on films useful to staff and volunteers of child-development programs. It also includes a listing of rental libraries throughout the country where many of the films are available free of charge and addresses of the major film distributors and suppliers of audio visual materials.

Pacific Oaks College, Pasadena, Calif. 91105. Publications list available.

Parent Cooperative Preschools International (PCPI), Whiteside Taylor Centre for Cooperative Education, 20551 Lakeshore Road, Baie d'Urfe, Quebec, Canada.

Play Schools Association, 120 West 57th St., New York, N.Y. 10019. Films play techniques for various settings, planning for excursions, and materials for use in parent education are available.

INDEX

NOTE: *Several references in this student edition refer to chapters found in the expanded, professional edition, and the reader is referred to it for the sources of these few references.*